Peter Allen

Lectures on Aural Catarrh

Or, the Commonest Forms of Deafness and Their Cure. 1872

Peter Allen

Lectures on Aural Catarrh
Or, the Commonest Forms of Deafness and Their Cure. 1872

ISBN/EAN: 9783337022242

Printed in Europe, USA, Canada, Australia, Japan

Cover: Foto ©berggeist007 / pixelio.de

More available books at **www.hansebooks.com**

LECTURES

ON

AURAL CATARRH;

OR THE COMMONEST FORMS OF DEAFNESS AND THEIR CURE.

(MOSTLY DELIVERED AT ST. MARY'S HOSPITAL.)

BY

PETER ALLEN, M.D.,

FELLOW OF THE ROYAL COLLEGE OF SURGEONS, EDINBURGH; MEMBER OF
THE ROYAL COLLEGE OF SURGEONS, ENGLAND; AURAL SURGEON TO,
AND LECTURER ON AURAL SURGERY AT, ST. MARY'S HOSPITAL;
AURAL SURGEON TO THE ROYAL SOCIETY OF MUSICIANS;
LATE SURGEON TO THE METROPOLITAN EAR INFIRMARY, SACKVILLE STREET.

NEW-YORK:
WILLIAM WOOD & CO., 27 GREAT JONES STREET.
1872.

TO

SAMUEL ARMSTRONG LANE, Esq.,

MEMBER OF THE COUNCIL AND COURT OF EXAMINERS OF THE ROYAL

COLLEGE OF SURGEONS IN ENGLAND;

SENIOR SURGEON TO ST. MARY'S HOSPITAL, ETC., ETC.,

These Lectures are Inscribed,

IN TESTIMONY OF RESPECTFUL APPRECIATION OF HIS EMINENCE

AS A SURGEON AND TEACHER;

AND OF GRATITUDE FOR KINDNESS RECEIVED,

BY HIS COLLEAGUE AND SINCERE FRIEND,

THE AUTHOR.

PREFACE.

There has not been, so far as I am aware, any new separate work, however limited in design, published in this country upon Aural Disease, since that of Mr. Toynbee, in the year 1860. It is obvious that the following Lectures do not pretend to be an exhaustive treatise upon their subject, which though it might at first sight appear to be confined within moderate limits, is in reality of very wide extent. I have, however, been anxious that this contribution to a most important department of study should have a certain completeness in itself, and should, whatever its many faults or deficiencies, embody in some degree the ideas fully established in my own mind, of the *system* essentially necessary to be pursued in striving to advance our scientific knowledge. This point (of system), appears to me to have been frequently lost sight of in some of the most esteemed and elaborate works on Aural Surgery and Medicine, with the consequence of (to say the least) impairing their usefulness to the young student. Nor has the busy practitioner the time to search and sift among endless records of cases, ranged under multitudinous classes, from the mass of which it is often a task of extreme difficulty to obtain any assured clue for his own guidance. I have, accordingly taken considerable pains in the construction of what may be called the ground-plan of the ensuing short series of Lectures; and not to be further tedious, I now conclude, by expressing my hope that they will be found by any earnest reader, of some genuine help, in telling him *how to examine, what to look for*, and *where to find the disease*, in a case of Aural Catarrh.

P. ALLEN.

15, SAVILE ROW,
    Dec. 30*th*, 1870.

CONTENTS.

LECTURE I.

INTRODUCTORY.—Subject of the lectures. Frequency with which aural catarrh originates in the throat. Parts affected in this disease. Neglect of aural surgery. St. Mary's Hospital the first to establish special instruction for students. Labours of Toynbee and others. Diseases of the external and internal ear not under consideration in these lectures. Middle ear, including the membrana tympani, the region most commonly affected in aural catarrh. Membrana tympani belongs to external as well as middle division of the ear. Importance of ocular inspection under all circumstances. Two cases in illustration. Remarks upon them . 1—17

LECTURE II.

Plan of examination adopted. Method of examining the auditory canal and membrana tympani. The aural speculum. Illumination. The concave mirror. Obstructions met with on inspecting the membrana tympani. Omissions in text-books with regard to the anatomical characters and appearances of the tympanal membrane *during life;* modes of testing the degrees of deafness. Physiology of listening. Reference to the stapedius muscle as the chief agent in *accommodating* the ear to minute sounds. Hearing better in a noise;—physiological explanation of it. Contraction of the tensor tympani muscle effects improvement in the hearing by its influence upon the relaxed membrana tympani, by reflex action. Analogy between functions of certain parts of the eye and the ear . . . . 18—35

LECTURE III.

Method of examination continued. Phenomena of the tuning-fork. Its great usefulness in distinguishing cases of so-called "nervous" deafness from others, and how it does so. Anatomy and appearances of the healthy membrana tympani in the living. Its relative position; its inclination; form; size; curvature; lustre; degree of transparency; texture. (Engraving of the membrana tympani as seen from without). Its structure, blood-vessels and nerves . 36—51

LECTURE IV.

The cavity of the tympanum. The ossicula auditûs. (Engraving of them and the membrana tympani seen from within.) Muscles of the ossicula. (Engraving showing the external and internal walls of the tympanic cavity, with the stapes.) Importance of the two fenestræ. (Diagram illustrating the relative positions of the various parts of the ear.) The walls of the tympanum described. Its mucous membrane. The Eustachian tube. Its anatomy. Its uses. The various effects of inflation of the tympanum through the Eustachian tubes. (Engraving of the tympano-manometer). Effects of deglutition, inflation, &c., as ascertained by the manometer . 52—69

LECTURE V.

On the different modes of examining and inflating the middle ear. Auscultation. (Engraving of the otoscope or diagnostic tube.) Valsalva's method of inflation. Politzer's, or the new method. (Engraving of the Politzer bag, with new nasal tube and pads attached.) Politzer and Valsalva methods compared; their relative value. Author's method of inflating the tympanum with his own Politzer bag. Politzer's method and catheterism compared. Catheterism of the Eustachian tube. History of the catheter. Mode of its introduction into the Eustachian tube described . . 70—87

LECTURE VI.

Catheterism, continued. Precautions to be observed. The air douche. Injection of fluids;—caution necessary. The various uses of the catheter. Effects of catheterism upon the middle ear. The sounds heard through the otoscope by the surgeon. Case of what the author terms *emphysema* of the membrana tympani . . 88—97

LECTURE VII.

Aural catarrh;—commencement of study of. Authors on this subject; their divisions appear unsatisfactory. Plan. Neither an anatomical nor a strictly pathological nomenclature quite feasible. Synoptical arrangement. 1st. Simple aural catarrh, subdivided into acute and chronic. 2nd. Purulent aural catarrh or otitis. 3rd. Otorrhœa, aural polypi, &c., or the results of the purulent form. Acute catarrhal inflammation. Degrees. Symptoms. Diagnosis. Myringitis, or acute catarrhal inflammation of the membrana tympani. Pathology of myringitis. Symptoms. Appearances of the membrana tympani and meatus externus in acute catarrh . . . . 98—112

LECTURE VIII.

Simple aural catarrh continued. Acute inflammation of the Eustachian tube, fauces, tonsils, &c. Symptoms. Diagnosis, Results, Prognosis, and Treatment of acute aural catarrh of the middle ear. Two cases . . . . . . 113—136

LECTURE IX.

Chronic aural catarrh. Pathology. Chronic catarrhal inflammation of the mucous membrane of the tympanum. Effusions. Exudations. Rigidity. Membranous bands. Anchylosis of the stapes to the fenestra ovalis. This result of chronic catarrh termed by the author, *impaction of the stapes*, in contra-distinction to Toynbee's *anchylosis* of ditto. Chronic catarrhal inflammation of the membrana tympani :—Appearances in . . . . 137—149

LECTURE X.

Chronic catarrh of the Eustachian tube. Sir William Wilde's opinion as to an impervious condition of it being a cause of deafness. Dr. Jago's investigations. Thickened mucous membrane the commonest and most immediate cause of obstruction of the tube. State of tonsils and fauces. Enlarged tonsils assist to produce deafness in other ways than by their direct pressure upon the orifices of the Eustachian tube. "Adenoid" growths, vegetations, or granulations cause defect in hearing. Reference to Dr. Meyer's paper on the subject. "Ulcerated sore throat." Cleft palate ; nasal and pharyngeal polypi ; fibrinous adhesion, and stricture of tube; relaxation of the mucous membrane of fauces; all more or less causes of deafness, by continuing their morbid effects into the Eustachian tube and cavity of the tympanum through the intervening mucous membrane. Chronic catarrhal inflammation of the mastoid cells. Symptoms of chronic aural catarrh. Tinnitus or "noises in the ear." Vertigo. Feeling of fulness in the head, heaviness, and various intra-auricular sounds. Tinnitus better understood than formerly . 150—172

LECTURE XI.

Chronic aural catarrh continued. Diagnosis and prognosis. Essential aid afforded by analogical anatomy, physiology, and pathology. Assistance to be derived from comparing respective offices of the component parts of the ear and of the eye. Some of these functions compared at length. Local treatment of chronic catarrh of the middle ear by German aural surgeons, also by the air-douche, vapours, and injections of fluids. Incision or paracentesis of the membrana tympani, too frequently performed in this country and in cases where contra-indicated. Other and less injurious modes of inducing absorption and dispersion of fluid accumulations and deposits suggested; proofs given of their success. (Engraving of syringes for the diffusion of medicated lotions as spray against the diseased mucous membrane of the throat and nose.) Gargles. Excision of tonsils; when desirable. Constitutional treatment. Four cases in illustration of the various modes of treating chronic aural catarrh . . . . 173—221

LECTURE XII.

Purulent aural catarrh or otitis. Its different forms. Often overlooked or not properly interpreted. Mr. Pilcher's opinions, &c., regarding the serious consequences of scarlet fever. Pathology. Symptoms and diagnosis. Infantile otitis especially considered, because often neglected. Terminations of purulent aural catarrh. Perforations. Otorrhœa. Abscess in mastoid cells. Cerebral abscesses. Caries of temporal bone. Diagnosis of perforations. Aural polypi. Granulations, &c. Treatment of otitis, otorrhœa, perforations, polypi, &c. On the artificial membrana tympani; history of its discovery by Dr. Yearsley. Mode of application, and highly valuable results. Two cases. Conclusion . . . . 222—271

EXPLANATION OF THE PLATE.

A Section of the nose, mouth, soft palate, uvula, tonsil, pharynx, upper part of the œsophagus, and trachea, showing the route taken by the Eustachian tube catheter along the inferior meatus and floor of the nostril into the orifice of the Eustachian tube: Showing also some of the parts of the throat which could be syringed with pulverized fluids by the elastic bottle and tube, illustrated on page 200.

ON

AURAL CATARRH,

&c., &c.

LECTURE I.

INTRODUCTORY.

GENTLEMEN,—The Course of Lectures on Aural Surgery given during the Summer Session being limited to twelve in number, it is impossible to notice within this period, except in a very cursory manner, many of the affections of the hearing organ which present themselves for treatment in the out-patient department of this hospital, under my superintendence. I purpose, therefore, in these lectures, to confine my observations mainly to that class of diseases which recent and more correct pathological investigation has enabled us to include under the generic and appropriate term of AURAL CATARRH. I shall also direct your attention to the origin and progress of some disorders, kindred in their nature to the above-mentioned, and which affect generally the same parts of the ear, although they may be found to ensue from specific causes, such as the exanthemata, rather than from ordinary ones. These ailments are often most serious in their results, and occasionally fatal in their termination. I may, therefore, with much reason, designate these lectures as being "ON THE COMMONEST FORMS OF DEAFNESS AND THEIR CURE."

In considering the morbid changes that occur in the auditory apparatus when affected by catarrh or by the exanthematous disorders (especially scarlet fever or measles), we shall find that the middle ear or tympanum is the part of the organ most implicated. This manner of dividing the subject will eliminate from our present design very many of the diseases of the external ear, comprising the auricle and meatus, as well as those of the internal ear, consisting of the nervous portion and labyrinth. Yet, even though we thus exclude all those disorders not arising from, nor complicated with, any abnormal condition of the middle ear, its contents, and appendages, it may still be safely asserted that in fully five-sixths of the cases which come under the notice of aural practitioners, the disease and accompanying deafness have originated in the mucous membrane lining the throat, nasal passages, and middle ear.

Before, however, entering further upon the actual subject of these lectures, a few words on the importance of the study of aural disease in general will not here be out of place. There are certain departments of professional knowledge, upon the practice in which the various Examining Boards exact no compulsory attendance, and seldom or never test the proficiency of candidates at the final examinations for diploma or licence,—and which, (as a natural consequence) obtain but scanty attention from medical students. Aural medicine and surgery is one of these subjects; its study is therefore deplorably neglected, and does not keep pace with that in other lines of scientific medical acquirement.

To remedy this acknowledged defect in the system of professional studies, the Governors of St. Mary's were the first hospital authorities enlightened and liberal enough to institute a separate department of aural surgery, and to provide special clinical instruction by means of regular courses of lectures on the subject. How ably the work was conducted by my predecessor, the late Mr. Toynbee, during the time he held the appointment of Aural Surgeon, is recognized and remembered by a numerous group of living practitioners, formerly students at this hospital. For several years he zealously laboured to elevate the study of aural surgery, which he pursued with so much success, to a position which, but for his indefatigable energy and industry, it would have been far longer in attaining. By his researches, and many contributions to our anatomical and physiological knowledge, as well as to the pathology of the subject (shown by the ample collection of specimens deposited in the Hunterian Museum), he immensely influenced the progress of aural surgery and acoustic science both in England and on the Continent. Aided by the microscope, and being accustomed to the work of minute dissection, pursued with patience, and carefully recorded observations, he accumulated an enormous mass of facts relating to the morbid anatomy of the hearing organ, which has formed the foundation of a more rational and successful system of treating the diseases of the ear than had before that time been adopted. His followers in Germany alone now number several distinguished aurists, whose theoretical teaching and scientific practice have given them a world-wide reputation. It would be invidious to specify names where all appear to be so devoted to their subject, are earnest workers, and seem animated by the desire to treat aural disease according to the true principles of modern medicine and surgery. They are determined upon extending the field of acoustic pathology in the best and surest way, that is, by carefully conducted *post-mortem* examinations, accompanied with honest clinical study and instruction.

The most frequent result of an aural disorder is, as you know, deafness, and it need scarcely be reiterated that impairment of hearing, in proportion to its intensity, becomes one of the most severe afflictions incident to mankind. To some it is even more disqualifying for either the duties or the enjoyments of life than the loss of sight. One of the two great highways by which the intellectual faculties are reached, is closed to the person thus afflicted. To the naturally intelligent, therefore, the loss of hearing will be more grievous than the deprivation of any of the other special senses.

Not only to persons of cultivated intellect, and those who have to earn their livelihood by the exercise of a profession or calling, and in the daily business of life, is the faculty of hearing a matter of vital importance, but also to those whose daily resource and comfort is social intercourse; for by any high degree of deafness, persons are debarred from holding other than a limited, tedious, and almost painful converse with their fellow creatures—and, deprived of hearing they are deprived of all external source of happiness.

Deafness occurring from any cause, in a very young child, entails also, as you will remember, dumbness, because having never heard sounds he cannot imitate them, nor even learn to speak. To his forlorn condition, therefore, all the foregoing reflections apply with added force. Our sympathies are naturally excited for the deaf mute, and his terrible infirmity demands the special attention of the aural surgeon, in whose power it sometimes lies to mitigate the severity of the affliction, even though he be unable to remove it entirely. A sense of the great responsibility attaching to him ought, therefore, to impel any conscientious medical man to devote earnest attention to the affections of the organ of hearing, and to disregard no opportunity of acquiring that knowledge which will enable him to act, if need be, with decision and effect. It is obvious that to the country practitioner, a competent acquaintance with ear disease is even more essential, inasmuch as thrown on his own resources, he cannot, as in the metropolis, call in the aid of superior talent and experience.

Owing to the great variety of the symptoms which we shall have presently to consider, we must endeavour to bring our rules of treatment out of the narrow limits of an exclusive specialism. In the human system, where there is such intimate connexion and mutual influence of parts, *each* part will be best understood and treated by him who has the clearest notions of the general economy. The surgeon who intends to practise aural surgery must not be educated merely as an aurist, because exclusive, or almost exclusive, attention to an individual portion of the animal structure causes a confinement of mental vision, like the bodily near-sightedness which mechanics contract by pouring over minute objects. We may say of the aurist, as Sir Wm. Lawrence of the oculist,—" All his habits lead to a separation and insulation of the organ; the part is detached from the system, treated by washes, drops, and ointments," and such inefficient trifling impedes the progress of aural surgery.

To make the subject of aural diseases acceptable or profitable to the student and general practitioner, it is requisite to simplify as far as possible the details of examination and method of treatment. I do not profess to give in these lectures anything approaching a complete description of the structure and functions of the ear, and shall only enter upon just so much of the anatomy and physiology of the parts under consideration as may relate to our immediate subject, "*The Commonest Forms of Deafness, and their Cure.*"

DIVISIONS OF THE EAR.

The ear admits of being correctly enough divided, according to function, into two parts, viz., a *conducting apparatus*, and a *fundamental* portion. The former is only necessary to the perfection of the sense of hearing, whereas the latter is so truly essential to the perception of sound, that were we to trace it through the whole animal series, we should find its representative in the lowest of the descending scale of creatures capable of enjoying the faculty of hearing even in its most limited extent. Anatomists, however, divide the ear into three parts, which I have before mentioned, the *external*, *middle*, and *internal* ear; and this appears to be the most natural and useful division, being in accordance with the relative positions of the parts, as well as convenient in considering the peculiar functions and disorders belonging to each.

1st. The EXTERNAL ear, which includes the auricle and external auditory canal.

2nd. The MIDDLE ear, being the tympanum with its contents, the ossicles, membranous partitions, the mastoid cells, and Eustachian tube. *Note.* These two divisions make up the "conducting apparatus" of the physiologist.

3rd. The INTERNAL ear, called from its complexity of structure, the LABYRINTH, consists of the vestibule, semicircular canals, and cochlea, and contains the expansions of the auditory nerve. This is the "fundamental" or essential part of the organ.

The foregoing division, both anatomical and physiological, being founded upon the comparative development in the animal series, is correct, and, moreover, eminently useful in studying the economy of the ear. It is found that not only is one portion superadded to the rest, according to the increased necessity of the sense, but that each department becomes more and more complicated until it reaches its perfection in the human ear. No further reference will be necessary in this course of lectures to the structure or diseases of the auricle or meatus auditorius, except so far as either is concerned in the extension of disease from the middle ear.

To prevent any misapprehension as to terms employed, I ought here, perhaps, to say that several writers place the membrana tympani (or drumhead) among the parts appertaining to the *external* division of the ear. The fact is that it cannot be considered to belong exclusively to either, but rather to both. Constituting the partition wall between the auditory canal and the cavity of the tympanum, it might well be reckoned in the two divisions, because it forms a part of each. The membrana tympani, as would naturally be inferred from its anatomical construction, participates in most of the diseases affecting the adjacent parts. The tissues upon either side are extended on its surface, viz., the epidermis and skin continued from the external auditory canal cover its outer side, while the mucous membrane, continued from the tympanic cavity, lines it on the inside; the proper fibrous structure of the membrana tympani being between them. It is also nourished by the same blood-vessels and nerves as supply the meatus and tympanum.

The anatomy of the membrana tympani should be well studied and known, because in it will be found characteristics of disease and changes from the normal condition sufficient to account for a very large proportion of cases of deafness. Sir William Wilde says fully three-fourths; others give a larger number. Having myself critically studied the history of cases detailed by all the eminent writers on aural medicine and surgery, and judging also from the results of my own practice, I have arrived at the conclusion that inflammations of the membrana tympani are very seldom uncomplicated. The appearances of the membrane, as revealed by inspection, furnish almost certain means of diagnosing the existence of morbid processes going on in the neighbouring structures. The "*myringitis*" of Wilde and Lincke (true inflammation of the membrana tympani) scarcely ever occurs as an independent disease.

Catarrh of the tympanum, whether acute or chronic, severe or mild, will almost certainly involve the membrane in its destructive advances sooner or later; and so will, likewise, a diffuse inflammation of the auditory canal. It is the intimate connexion of all these parts, as before mentioned, with the important membrane of the drum, which causes morbid changes originating in the outer or middle ear to be transmitted to it. So that, in speaking to you of the disorders and functional disturbances produced by aural catarrh, and of the commonest forms of deafness, I shall over and over again refer to appearances presented by the membrana tympani when submitted to proper inspection.

It is true that we sometimes do not see indications of any changes on the external surface of the membrane, sufficient to account for the high degree of deafness which may exist; and even occasionally, no departure whatever from the normal condition can be perceived upon it. Although such a negative result of inspection does not possess the value which the discovery of marked signs of disease, past or present, would do, yet it must not be underrated; for, by joining it with the history of the case and the subjective symptoms as given by the patient, we are enabled to exclude from our diagnosis any affection of the tympanum, and we correctly infer that the cause of the impairment of hearing lies in the deeper structures of the ear, probably in the fenestræ of the labyrinthine wall, or in the nervous expansion within the labyrinth itself. Thus, an ocular inspection, carefully made, will not only satisfy us as to the natural colour, form, position, transparency, curvature, &c., of the membrana tympani, and the existence or otherwise of any morbid processes in and upon it, but will also contribute, in conjunction with other tests and symptoms, to an exhaustive diagnosis. In fact, whether by a close and searching inspection we discover unmistakeable signs of disease, or only a perfectly natural state of the membrane, we must have recourse to it; we thereby are helped to localise the affection which causes the functional disturbance, and do not fruitlessly apply our treatment to the wrong part of the organ.

The following is an instructive case, well illustrating the importance of a careful examination of the auditory canal and membrane of the drum, and also the mistakes which occur from neglecting it.

A Case of supposed nervous and incurable Deafness, attended with pain, giddiness, and slight cerebral symptoms, ascertained by inspection to be caused by an impacted layer of hardened cerumen pressing against the Membrana Tympani. Cured at once by the use of the syringe.

Mrs. M., æt. fifty, whose husband was consulting Mr. Erasmus Wilson, was advised by that gentleman to apply to me (in August, 1869), on account of severe deafness on the right side. It was considered by the physicians and surgeons whom she had before consulted, to be "nervous and incurable." Her sister, and several cousins were somewhat deaf also, a circumstance which gave rise to the idea that the infirmity must be "in the family." She first perceived her hearing to be impaired about two years ago, since which time the deafness had increased, until at the time of my seeing her, she could not distinguish the tick of my watch (a powerful one, usually heard at ten or twelve paces off), except in contact with her ear. She required to be spoken to in a loud voice, and within a few feet distance. Singing or tinnitus was occasionally distressing, and sometimes there was dizziness. The most alarming symptoms, however, were paroxysms of severe pain in and around the ear, extending to the top of the temporal region, and downwards to the jaw, accompanied with vertigo, to relieve which her medical advisers had recommended, from time to time, leeches and blisters alternately.

This treatment gave temporary relief to the pain, but in no degree mitigated the deafness. On placing the tuning-fork on the vertex of the head, for the purpose of testing the condition of the auditory nerve (I shall refer in subsequent lectures to the value of this instrument in examinations), it was heard best on the right side—in the deaf ear. This satisfied me that the patient's deafness was not "nervous," as previously imagined, but that some impediment to the conduction of sound *out* of the ear, as well as *into* it, existed. The Eustachian tube was pervious, but inflation of the drum by the Valsalva method produced pain in the affected ear. The sound, too, of the compressed air against the membrana tympani, as ascertained by auscultation with the otoscope, was muffled, and the inflation failed to give that decided flapping or crackling sound which usually follows the forcible pressure of air into the tympanic cavity. By the patient's own sensations under this diagnostic experiment, and by the absence of any special indication of aural catarrh, except so far as the vibratility of the drum was interfered with, I felt convinced that I must look for some cause which prevented the free entrance of sonorous vibrations into the ear, and that the hindrance would be found to exist in the meatus or in the membrana tympani. Accordingly, when the passage of the outer ear was well illuminated (in a manner I shall presently define), a darkish, thin substance appeared, spread over the surface of the membrane. This was, with caution and some little difficulty, removed entire, by syringing with warm water. Instantaneous relief followed, as well to the deafness as to the feeling of oppression and discomfort. The obstruction thus brought away proved to be nothing more than a thin scale of very hardened cerumen, resembling flattened vulcanised India-rubber, which, being closely impacted against the membrana tympani, took its shape and curvature, fitting into the external concave surface as if plastered on it, and forming a cast of it. Of course, the delicate vibratory motion of the membrane was thus destroyed, hence the high degree of deafness.

The constant pressure upon the vascular network of the dermoid layer of the meatus (which layer is continued, as before remarked, around and on the membrana tympani), as well as upon the numerous nerve filaments accompanying the arteries, caused the cerebral congestion and paroxysms of pain; and, lastly, the ossicula auditûs, which you will recollect form a chain extending from the inner surface of the membrana tympani to the wall of the vestibule, being unduly pressed against the labyrinth-fluid, the alarming attacks of giddiness were accounted for.

The case above detailed, not only affords a practical example of the necessity of closely inspecting the outer surface of the membrana tympani and the meatus, to enable us to form a conclusion regarding the state of the outer and middle ear; but it also confirms a fact to which I may rather frequently allude when speaking of the symptoms of aural catarrh;

that superficial inflammations of the membrane, and undue pressure upon its exterior are always painful, and sometimes occasion cerebral disturbance, while extensive changes may occur on the inner, or mucous layer, without producing the slightest pain or "earache," from which children so frequently suffer.

As the converse of the above instance, where mechanical impediment so manifestly caused the deafness, by preventing vibrations from reaching the middle and internal ear, and gave rise to the painful symptoms, I will now relate some particulars of a case where careful ocular inspection afforded only the slightest evidences of an aural disorder, though the impairment of hearing was quite as great.

Yet, in spite of the absence of the usual characteristics of disease in the following case (the result of inspection being in fact almost negative), my examination led to the valuable inference that the cause of the extreme deafness must be located in the parts contiguous to the membrane, which are associated with it in function.

Case of Aural Catarrh with obstructed Eustachian tubes, producing considerable deafness, which was supposed by the patient to be caused by an accumulation of cerumen in the Ear.—Cure.

C. J. P., Esq., a country gentleman, consulted me in January last. On coming into the room, he said, "I believe both my ears are stuffed up with wax. I am very deaf." The *history* was that, about six weeks previously he suffered from a bad cold, and found himself gradually becoming deaf. He had no pain, but there was a constant singing and beating noise which was very unpleasant, and a sensation of fulness and weight on both sides of the head and in the ears. The deafness did not vary at different parts of the day, and he had to be spoken to in a loud tone of voice. He could only hear my powerful watch half an inch from either ear. The tuning fork placed on the vertex was distinctly and equally heard on both sides. After looking at and examining in the way presently to be described, both the auditory meatus and membrana tympani, I informed him that there was no unusual secretion of cerumen whatever, at which his countenance fell, and he exclaimed, "What, must I be deaf for life? Is there no remedy?" The result of my examination, however, permitted me at once, to state confidently that a very simple kind of treatment would probably restore his hearing. Now, the only appearance of irregularity which I could make out, was an exaggerated concavity of the membrana tympani in both ears, their surface being also slightly opaque and dull-looking. I placed the otoscope in the meatus, and directed the patient to make a strong attempt at expiration, while the mouth and nostrils were tightly closed. With scarcely any difficulty he succeeded in forcing a stream of air through the Eustachian tubes into the tympanum, and the loud crackling sound produced by the rapid bending outwards of the drum-head, reached me distinctly through the otoscope. So complete and sudden was the inflation, that he compared the feeling to that of a gun going off in his head, and he exclaimed a moment afterwards, "I can hear your watch tick in your pocket!" The annoying tinnitus or singing ceased, and my watch was heard plainly at the end of the room, a distance of twenty-five feet from where he was sitting. The mucous membrane of the fauces being rather relaxed and granular, an astringent gargle was prescribed, which restored it to a healthy condition, and my patient reported that he had "no return of deafness, and the air passed into the ears quite well." Many similar cases, you see treated in the out-patient's room with just as satisfactory results.

This case, you will perceive, differs from the preceding one, in some important particulars, which I will specify as concisely as possible,—ocular inspection and auscultation having been alike imperatively necessary in both instances. In the former case, the tuning fork and successful inflation of the drum, assured me that the lesion causing the deafness was not in the internal or in the middle divisions of the ear, but that it was to be sought for in the external ear or surface of the membrana tympani. Inspection not only confirmed the diagnosis in these respects, but revealed the actual presence of the small plug of indurated wax which was the cause of all the mischief and distressing symptoms. In the latter case, inspection of the membrana tympani materially facilitated the diagnosis, by showing me only a very slight departure from its normal condition, but which a practical knowledge of its appearance under variable influences, and irregularities of function, enabled me to associate with certain disorders of the middle ear or its appendages; while the tuning fork assured me of the perfect action of the auditory nerve expansion on the inner division of the ear. In both cases the state of the membrana tympani, ascertained by direct ocular examination, readily explained the nature of the disease which was causing so much functional derangement, and guided me to adopt treatment which speedily cured the patients. Once more, then, let me reiterate, that we must never omit ocular inspection in examining our patients, and as our methods of exploring the dark passages of the ear become more perfect, so will our perception of the characteristic signs of disease be made easier, and become of more diagnostic value, while our treatment will be more scientifically and successfully applied.

The difficulties attending a satisfactory inspection of the ear will be for the most part overcome by practical experience in using the speculum; whereby the variable aspects of the all-important membrana tympani, both in health and disease, will be recognised and determined. Furthermore, certain appearances, or deviations from the normal state of the membrane which the practitioner may observe, will be gradually and naturally associated in his mind with certain corresponding functional disorders of the parts seen, or even of those beyond the reach of the eye. I shall, therefore, next proceed to give you a somewhat detailed description of the methods of examining an aural patient.

LECTURE II.

Gentlemen,—As the following plan which I pursue in eliciting all the important particulars of an aural case that are necessary to assist me in forming a diagnosis, appears to possess some advantages over others in vogue, I refer to it in a few lines. The subjective symptoms being usually given in the patient's own words, it saves time to take them down in a categorical order. The list of questions which I require answered is printed in marginal lines in my note-book, with blank spaces for filling in. Thus—

    Name
    Residence
    Age
    Date
    Ear affected
    Ear most affected
    Presumed Cause

PLAN OF EXAMINATION.

Predisposition
State of Health
Prior Treatment
Period of Disease   Right   Left
Tinnitus   Right   Left
Pain   Right   Left
Discharge   Right   Left
Hearing Distance   Right   Left
Meatus Ext.   Right   Left
Memb. Tymp.   Right   Left
Eustachian Tubes   Right   Left
Pharynx
Nasal Passages
Auditory Nerve   On which side tuning fork is best heard.
Diagnosis

Treatment and Remarks

The progress of the case can be carefully watched, and the hearing distance noted down at each visit. By this latter means we oftentimes obtain unmistakeable proof that our patient's improvement is more satisfactory than he is inclined to believe.

Omitting for the present any further allusion to the subjective symptoms, which the patients themselves furnish us with, I now go on to demonstrate to you the best modes of examining those parts of the hearing organ which are implicated in the commonest forms of deafness,—Aural Catarrh, the immediate subject of these lectures. The term is used for the sake of brevity. Aural catarrh must not, therefore, be considered as an affection in which the mucous membrane of the cavity of the tympanum and Eustachian tube is solely involved, but it is applied also to catarrhal inflammations of those structures which are lined with a continuation of the mucous membrane of the tympanum, as well as to certain chronic forms of inflammation of the dermoid layer of the meatus. It is convenient to designate this most frequent disorder of childhood as a catarrhal affection. It constitutes the "ear-ache," and, if treated lightly, results in the establishment of an offensive discharge or otorrhœa, which impairs the hearing, and may, by extension inwards, set up diseases which have a fatal termination. An aural catarrh, originating in the mucous membrane of the cavitas tympani, or in the dermoid layers of the meatus, will, sooner or later, furnish unmistakeable signs of its presence on the membrana tympani; and we must, in order to recognise the appearances which are indicative of it, critically examine the condition of the auditory canal and the drum-head at its extremity,—the all-important, vibrating, elastic, semi-transparent boundary wall of the tympanum. Let me here emphasize the caution not to confound the terms "drum-head" (membrane of the tympanum) with "drum" (cavity of the tympanum), as is done by most of the German writers on aural surgery, to the constant misleading of those readers who are not familiar with acoustic subjects.

METHOD OF EXAMINING THE AUDITORY CANAL AND MEMBRANA TYMPANI.

The Speculum.—A few years ago, when the best text books on aural surgery were written by Pilcher, Toynbee, Yearsley, and Wilde of Dublin (of whom the last is now the only survivor), examinations of the parts in question were most unsatisfactorily made, because the instruments then employed for the purpose were altogether inadequate for enabling them to obtain distinct views of those minute alterations on the membrana tympani which are now, by recent and improved means, rendered perceptible. The valvular or forceps speculum auris of Kramer, which was intended to dilate the cartilaginous portion of the meatus, is now seldom used. It is a cumbrous, heavy instrument, requiring to be held in the meatus with one hand, while, if any operation has to be performed, the other hand only is available to do it. The usual growth of hair in the passage, loose epidermis, or cerumen may, by being projected between the blades, quite obstruct the view of parts beyond. The ear specula, here shown, which I use and find to answer all my requirements, are tubular and of various sizes. Some are made of silver, others of porcelain or glass. The smaller extremities are either circular or oval, and with or without indentations for their more ready passage over the dermoid layer of the meatus, which, in certain states, is apt to become raised and pushed before the instrument. The inventor of the india-rubber specula, Dr. Politzer, of Vienna, truly claims for them the advantage of being lighter than the others, and they will consequently remain in the meatus without support. The dark ground of the inner surface favours the distinctness of the parts illuminated, and the blunt edges prevent abrasion of the lining membrane.

Illumination.—Until quite recently we have been accustomed to use direct sun-light, or strong artificial light, to illuminate the external surfaces of the meatus and membrana tympani, but both are, for the following reasons, objectionable:—The surgeon had to bring his patient close to the window, so as, if possible, to get the rays thrown into the passage through the speculum. Patients sometimes could not be moved from their beds, nor will the pleasant beams of the sun always favour us with their presence in this our variable climate. The surgeon, too, in using direct light, must stand between it and the patient, and finds it extremely difficult to avoid throwing the shadow of his own head, and so precluding all possibility of seeing far into the speculum. Sunlight is glaring and dazzling to the eyes of the observer, and there is the same objection to its employment as to strong artificial illumination by means of lamps. The former, by its brilliancy, impairs the distinctness of an object, and the latter has the effect of altering the colour of the membrane, giving it more or less an unnatural appearance, and preventing minute changes on its surface from being duly recognised.

The disadvantages of these modes of illumination are now overcome by the use of a concave mirror, with five to six inches focal distance, which enables us to reflect powerful beams of daylight or artificial light, through the speculum, into the remotest part of the canal, thus obtaining a perfectly defined illuminated view of the whole surface of the membrana tympani. The patient under examination is not placed in a constrained posture, as he must necessarily be when direct light is the only illuminating means employed; the head may be turned in any required direction, and, if he is seated in a revolving chair, we do not oblige him to rise when we require to look into each ear by turns, perhaps several times during the course of the consultation. Ordinary diffused daylight is, with this contrivance, to be preferred to sun or artificial light, because it does not change the colour, nor in the slightest degree alter the appearance of the structure inspected.

The simple methods above detailed have now superseded all others. During operations, such as the removal of polypi, incisions, the application of escharotics, or when carefully observing the varying aspect of the membrane of the tympanum while the latter is subjected to inflation through the Eustachian tube; the mirror (being also provided with a ball and socket joint) may be fastened to an elastic band, which is passed round the head of the surgeon. This has the great advantage of leaving both hands at liberty. The patient's head being placed nearly on a level with, but somewhat inclined away from, that of the surgeon, who holds the mirror in his right hand, or affixed to his head, the selected speculum is introduced into the meatus, and gently pressed inwards by the left hand with a slight rotary movement. The speculum may now be moved in any direction, while the different portions of the auditory canal and surface of the tympanal membrane are brought into the field of vision. I have here, also, a little apparatus for exploring the outer division of the ear, which, with some trifling modifications, may, I hope, enable us to obtain a clear view of the accessible parts of the membrana tympani under magnifying power. I require, however, further experience in its use before recommending it.

The practitioner will frequently meet with *obstructions* in the external meatus, sometimes found in the cartilaginous and sometimes in the osseous portion, which intercept the perfect view of the membrane beyond. Over-abundant secretion of cerumen, or impaction of the same, an unusual growth of hair, or detached epidermic scales, are obstacles requiring careful removal before we can obtain a satisfactory sight of the parts. A more formidable species of obstruction, a too great curvature of the walls of the meatus, (generally the anterior and lower) is occasionally present. Morbid processes in the meatus, inflammations of its dermoid layer, causing constriction of the passage, purulent accumulations, granulations, polypi, or exostoses, render the membrane inaccessible to inspection.

However, supposing we have succeeded in bringing

clearly and distinctly into view all the parts we desire to examine, we must now ask ourselves the question—Do any signs, characteristic of disease, present themselves on the membrana tympani, or are we looking upon a perfect and healthy structure?

The anatomical characters of the membrane, when in a normal state, are, unfortunately, but little familiar to students or practitioners, and even our most efficient text-books fail to give us trustworthy information on the many most important points of its structural condition, which have an especial bearing upon its pathology.

I need scarcely reiterate that to be intimately acquainted with the various appearances of the membrana tympani *in health*, as revealed by the recent improved methods of inspection, and to be able to recognize minute changes and departures from the normal state, when diseased, is an acquirement of the first value to the practising aural surgeon. Moreover, when by carefully recorded observations made upon aural patients *during life*, compared with corresponding appearances *after death*, he has established, so to speak, a perfect harmony between conjectural and verified causes of disease, he possesses such a comprehensive knowledge of his subject as will render errors in diagnosis almost impossible; and he will thus be prepared to treat successfully, in their earlier stages at least, the vast majority of cases which have heretofore been allowed to go on to incurable deafness.

As these most desirable results, however, are only

to be attained by studying and comparing the phenomena presented in health as well as in disease, I shall consider it necessary to bring before you, in a condensed form, the anatomy of the membrana tympani (which, as I have before said, stands in such very intimate relation with all the catarrhal diseases of the middle ear), so that subsequently we may the better understand its pathology.

In the anatomical detail you will find we shall collect together all the most prominent points which are noticed both in health and disease when we look into the auditory canal, and, as I said before, bring the membrana tympani clearly and distinctly into view. We shall then carefully consider its colour, degree of transparency or opacity, its tension, lustre, curvature, and its inclination or position with regard to the interior of the cavity, of which it forms the outer boundary, the direction and situation of the handle and short process of the malleus. Not, however, to unnecessarily introduce confusion into the subject which, at present, more directly engages our attention, viz., the methods of examination. I think this is the most appropriate time to speak of the manner in which we ascertain the *exact degree of deafness* under which our patients labour.

In measuring the distance at which persons can hear the watch, the surgeon should bring it gradually to the ear and withdraw it in the same way; and a written note of the ascertained distance ought always to be taken. The watch is not a perfect instrument for this purpose; but notwithstanding its insufficiency on account of its possessing but one or two tones, which convey only separate shocks to the auditory nerve, we have no better or more convenient appliance. We must also make inquiry as to the power of hearing conversation, and whether high or low tones are best appreciated. Some persons hear perfectly well when they are listening, but as soon as the voluntary act is suspended, a conversation maintained in the same tone of voice is not perceived by them. It is thus that the friends of persons afflicted in this manner sometimes imagine that there is no defect in hearing, but merely a want of attention. The intelligent surgeon will in such a case recognize an impairment of function in certain parts of the sound-conducting apparatus; and his knowledge of aural physiology will guide him to associate such a condition with derangement of some structure which is influenced by a strong effort of the will.

Instances like the above not infrequently present themselves; and, to make this matter clear to those of you who may not yet be sufficiently advanced in a knowledge of the nervous system, let me explain to you that the two muscles which move the little bones of the ear (*ossicula auditûs*), and regulate the tension of the membranes, receive their nervous supply from two different sources—one being under the control of the will, and the other not. When, therefore, in testing the degree of hearing, we discover an inability on the part of our patients to distinctly hear words addressed to them unless they are looking at you, and bestowing earnest attention on what is said, it may be concluded that something impedes the action of the *voluntary* muscle of the ear, not the involuntary. Now, the muscle (*stapedius muscle*) which draws the stapes obliquely out of the fenestra ovalis, and thereby relaxes the membranes, enabling them to become impressed with most delicate vibrations, is supplied with a nerve from the portio dura (the facial of the seventh pair), which is a voluntary motor nerve. Reverting from effect to cause, we find that when the mucous membrane lining the walls of the drum and its ossicles becomes congested, thickened, and rigid, it binds down, more or less strongly, the stapes to the fenestral opening, and restricts its movements. Hence a vigorous and sustained action of the stapedius muscle is required, to pull out the little bone, and overcome the counteracting pressure of the diseased mucous membrane, as well as the rigidity of the ossicles. Thus, then, it becomes evident to you how the inability to hear conversation without attentive listening on the part of the patient, will indicate a catarrhal affection of the cavity of the tympanum. This, therefore, should always be tested.

On the other hand, persons suffering from some kinds of deafness resulting from aural catarrh, hear considerably better, or well, when riding in a railway carriage, in an omnibus, amidst great noises, or loud rumbling, roaring sounds. A case is recorded where a man could only converse with his wife while a servant beat a drum. In another instance, a shoemaker's son could understand what was said to him only when he stood near his father pounding leather on the lapstone, so that when the latter wished to speak to his son, he took the hammer and pounded away at the leather. This boy could also hear well inside a mill, but outside of it voices were unintelligible.

In testing patients' hearing powers by means of the metronome (a most useful instrument, which I will show you after lecture), I find some are quite unable to distinguish the acute sound of the bell inside it when struck by the hammer, while others hear it more distinctly than I do myself. Such persons as the latter, however, cannot hear proportionately well the dull low sounds produced by the striking on the wood.

The immense improvement in the hearing which in some cases attends the patient's travelling in a carriage, or amidst the vibrations of any substance emitting low grave sounds, has been attempted to be explained in various ways, but none of these are satisfactory in my opinion. Some writers, even of note, consider that the ability to hear under these circumstances is imaginary, not real, or that, as Dr. Von Tröltsch says, " misapprehension and lack of observation are at the bottom of these statements." Nevertheless, as Sir William Wilde observes in his admirable treatise on aural surgery (when speaking of the symptoms in a condition of the drum head, termed by him " collapse of the membrana tympani"), " it is a well-established fact that certain deaf individuals will be able to hear the human voice in its ordinary tones, and to enter into conversation while travelling in a carriage, walking in a street through which vehicles are passing, or under any circumstances in which the air is agitated by sounds much louder than those in which the conversation they are listening to is addressed."

He then proceeds to mention that he knew an instance of a miller, who, like the boy above referred to, could hear perfectly ordinary conversation while standing within the working mill, but as soon as the mill ceased, or he removed into another locality, he could only hear when spoken to in a much louder tone of voice. This author, who does not, like some others, when unable to account for a peculiar phenomenon, evade the point, or doubt the assertions of the patients, candidly admits that no satisfactory explanation has yet been given.

In reference to Kramer's opinion on these cases, that " the auditory nerve becomes *so excited* by these deep-toned uniform noises whilst they continue, that the patient often hears the human voice better than a sound person whose ear is stunned by the noise," Sir William Wilde says:—" I admit the facts are certainly as thus stated, but the inference does not follow." He appears to attribute this singular advantage possessed by some deaf people over their sounder travelling companions to " the state of the membrana tympani," but does not explain how that peculiar condition should generally be associated with a power of hearing better in loud noises. He is assuredly, however, in error when he states further on, " that we do not witness these phenomena in cases where the membrana tympani has been in whole or in part removed." My own daily experience in the treatment of perforations, or in cases where a considerable portion of the membrane has been lost, leads me to an entirely different opinion.

While writing the notes for this lecture, I am called upon to apply the little plug of cotton wool (Dr. Yearsley's valuable invention, which he truly called from its effects the "magical" artificial tympanum), to a very deaf patient. She not only hears well in her carriage, &c., but I am able to make the tones of the metronome appear to her higher or lower, louder or duller, according as I exert pressure upon different parts of the largely perforated and otherwise diseased membrana tympani. This patient occasionally hears my watch at twenty-five feet distance, and her case will be hereafter again referred to.

You may yourselves generally see patients under my care here, from whom you should always elicit this peculiar symptom when they have it. But I wish specially to impress upon you that you must not, on any account whatever, assume that these symptoms we have been speaking of are characteristic of true *nervous* deafness, merely because they have not as yet been satisfactorily interpreted.

From careful study of the effects produced on my own hearing by the voluntary inflation of the tympanum through the Eustachian tube (by which the air in the drum becomes condensed, and its membrane stretched outwards), and also by an opposite state of the drum, when the air in it is rarefied or exhausted by being drawn out of it, I am convinced that I thereby change the character of certain sounds which fall upon the membrana tympani. The low grave tones of the metronome become raised in pitch, while the very high ones of the bell are not sensibly made more audible than before. It is possible, however, that they may be heightened, but beyond my perception. By this experiment, which you can make trial of while the metronome is before you, it is evident that when the membrana tympani is rendered tense by inflation or exhaustion, driving it either unnaturally outwards, or drawing it inwards, I am able to perceive sounds which did not appear to be yielded by the instrument before. Now, the loudness or intensity of the sound emitted by the metronome is the same in both instances, so that the alteration in pitch must be due to the increased tension of the membrane of the drum. From the results thus obtained by experimenting on a healthy membrane, it will be readily admitted that certain states of disease may so alter its functional capacity, as to render it incompetent to transfer across the tympanum those vibrations which impinge upon it. The disease may be such as to make the membrane incapable of vibrating at all, or it may affect it in a more limited degree, by which only *the rate* of its vibrations will be altered (by being diminished), and the *grave* tones (those having few vibrations) only perceived.

Stretched membranes, you must always recollect, take up vibrations from the air with great readiness. The preceding observations will have prepared you for the statement that the principal function of the ossicular chain and its muscles is to regulate the amount of undulations which are to pass to the nerve in the labyrinth, and we have seen that the tympanic membranes will vibrate according to their degree of tension or relaxation. It is probable that the membrana tympani is generally in a moderately relaxed condition, or scarcely on the stretch ; so that the tones of ordinary conversation (or somewhat grave ones) are the best suited for its perfect vibrations. Obviously, the various changes in the membranes cannot be effected without muscular action. The two muscles of the tympanum (*the tensor tympani and the stapedius*) by their action, combined or separately, regulate the tension of the various membranes that are made to vibrate, and thus, in conjunction with the ossicles, they act as the analogue of the iris in the eye. The tensor tympani primarily and chiefly influences the drum-head, by pulling inwards the handle of the malleus and the membrane in which it is embedded ; and, therefore, when contracting, it tightens the membrana tympani, just as the circular muscular fibres of the iris, by contracting, diminish the size of the pupil, and protect the optic nerve from receiving too powerful or destructive impressions of light.

Now, when patients are suffering from the effects of a catarrhal disorder which has left the drum-head in a relaxed, flabby, sunken, atrophied, thinned, or collapsed state, or in fact, in *any* condition which has greatly reduced its vibratility, we find that an improvement in their hearing takes place whenever *tension* of the diseased membrane can be accomplished. But, it will be asked, how can the deaf person, while riding in a railway carriage, or amidst a tumult of sounds and noises, as before mentioned, have the state of the membrane suddenly changed from relaxation to tension? The reply to this inquiry will, I am persuaded, furnish the true explanation of this curious fact. The tensor tympani, which alone tightens the membrane, receives its nervous supply from the otic ganglion, and from the motor root of the trigeminus, the third or inferior maxillary branch of the fifth. Its contractions are therefore, mostly if not entirely involuntary movements, unlike in this respect, as we have seen, to the stapedius. Politzer has, by irritation of the trigeminus within the cranium of a vivisected dog, stimulated the tensor tympani to action, and rendered the effect visible on the membrana tympani. This muscle is excited by reflex action (again like the iris), which may be caused, either by the expectation of a loud sound, or by any actual sonorous impression.*

To keep to our analogy : as the optic nerve, spread out on the retina, requires the stimulus of light to cause contraction of the muscular fibres of the iris, so does the auditory nerve need the stimulus of sound to put the tensor tympani into action. Now, the aforesaid vibrations of carriages, drums, leather poundings, mill grindings, or what not, being conveyed to the nerve of hearing, are either of them stimulus sufficient to bring about a reflex contraction of the tensor tympani muscle.

Our patient with the relaxed condition of membrane above described, has the tension of it effected

---

* The membrana tympani itself is likewise supplied plentifully with nervous filaments from the very compound fifth nerve—its inferior maxillary branch—and as *all* reflex actions are essentially performed independently of the will, it is almost certain that the *reflected* contraction of the tensor tympani muscle is involuntary.

by the very sounds which confuse us. Increased tension of our healthy drum-heads will prevent our reception of *ordinary* conversational sounds, whereas the relaxed drum-head of the patient, by thus becoming stretched and tense, is rendered more like our own, and is then in a state to receive and transfer sounds that were before inaudible. Simultaneously, a change occurs in the position of the ossicles, the whole chain is, as it were, braced up, and due tension of the inner membrane and pressure upon the fluid in the labyrinth effected. The patient, previously deaf, thus hears as well as, or better than we do ourselves under the circumstances specified. Reverting again from effect to cause, we see how diseases of particular portions of the tympanum are indicated by the symptoms, and these latter are to be discovered by properly directed tests and questions.

LECTURE III.

METHOD OF EXAMINATION—(*continued*).

GENTLEMEN,—You have seen, in the out-patient's room, what very valuable assistance to our means of diagnosis is afforded by the use of the vibrating *tuning-fork* placed on the vertex of the patient's head; in fact, you never witness a thorough examination of a new comer without this test being applied.

It is only within a short period that a knowledge of the fact of sound-conduction by the bones of the head has been rendered so practically valuable in the diagnosis and prognosis of aural disease, and we are chiefly indebted to Lucæ, Politzer, and Mach for their scientific investigations in this direction. The tuning-fork is employed principally in distinguishing affections of the *conducting* apparatus (the external and middle divisions of the ear) from those of the *auditory nerve*. It had been known for years that the watch or tuning-fork, placed on the forehead or between the teeth, was better heard if the auditory meatus was stopped with the fingers; but no very satisfactory conclusions were drawn, nor any material aid in diagnosis derived from this fact.

It must always be remembered that any sonorous vibrations which once get into the tympanum become intensified by resonance. The common speaking-tube is a familiar example of sounds being strengthened thus when confined in cavities of any sort. That this is the case with regard to the tympanum and its continuous tube, the osseous meatus, may be proved by closing the external passage with the fingers, when, if a tuning-fork be set vibrating on the head, or a humming sound or reading be kept up, the sounds, being conveyed through the cranial bones to the cavities of the ear, will become considerably intensified. This fact is made still more evident by placing a vibrating tuning-fork on the forehead, and stopping up one ear with the fingers; the sound will then be more audible on that side. Also, if the sounds of the fork have died away when listened to with the ears unstopped, they may be instantly restored, or rather, the perception of them revived, for a short time, by closing the ears. The way, therefore, in which we distinguish affections of the *sound-conducting* portions of the ear from those of the nervous (or *sound-perceiving*) apparatus is as follows:—If the patient be deaf to the sounds of a watch or a tuning-fork held near (not touching) the external meatus, and yet can hear distinctly their vibrations when conveyed through the solid structures of the head, teeth and the like, it may be inferred that some obstruction exists to the passage of sound through the meatus, membrana tympani, or tympanic cavity, but that the functions of the acoustic nerve are unimpaired. The surgeon may also assume that the conducting apparatus is in fault if the vibrations of the tuning-fork and the patient's own voice are not better heard when he closes his ears; because it has been shown by the above experiment that the closure of the meatus amplifies all sounds transmitted through the skull or interior of the mouth. It is obvious that catarrhal disease, whether of the tympanum, its contents, or its external membrane, would hinder the escape of the intensified sounds outwards through the meatus, just as effectually as would be done by a plug of cerumen or the stopping fingers. Patients are often inclined to state that they perceive the sounds to be more audible on the side where they can best hear conversation. This is because they naturally expect it must be so; but when we explain to them the method, and its important bearing upon the case, we find (should the disease be really in the conducting parts) that they will verify the correctness of our investigation by referring the increased perception of sound to the deaf ear, unexpected and surprising to them though this may be.

Consequently, you may generally and safely conclude that you have to deal with a case of obstruction to the free entrance of sound into the internal ear, and not with a nervous affection, if the patient admits that he decidedly hears the vibrating tuning-fork or watch on the deaf or deafer side. Inversely, of course, if the fork be heard very indistinctly or not at all when placed on the vertex, we must infer that the auditory nerve is not so sensitive to the impression of sounds as it ought to be, and that either there exists some abnormal pressure upon the labyrinth-fluid, or that the nerve itself is implicated in disease. The history of the case and further analysis of the symptoms will point out whether this condition is referable to a primary nervous affection, or a secondary one originating in the cavity of the tympanum. This subject will be further dwelt upon as we go on.

Inaccuracies which may occur in the patient's description of the effects of this experimental test can be corrected in a great measure by using another kind of otoscope or diagnostic tube. This instrument has two india-rubber tubes, about two feet long, joined to a single tube one foot in length, something like Dr. Scott Alison's double stethoscope. The two tubes are firmly placed in both auditory canals of the patient, and the common single one in the surgeon's ear. Now, if the tuning-fork be made to vibrate strongly and then placed on the vertex of the patient, sounds ought to pass equally out of both ears if there is no obstruction; and we ascertain whether they are thus freely transmitted on both sides, by pressing alternately the arm of each tube inserted into the patient's ears. If we find a difference in the intensity or clearness of the sound as conveyed to our own ear, we can determine through which ear of the patient the vibrations are most distinctly proceeding. Of course, the patient will, for the reasons before referred to, hear them best on the deaf side when there exists any impediment to their passage outwards, while the *surgeon* will hear them best through that tube connected with the patient's healthy ear, where there is no such hindrance to the exit of sounds. Very slight variations will probably not be detected by this means. Lastly, ascertain also whether the patient can hear the vibrations of the tuning-fork on the head for as long a time as you yourself can. The moment he ceases to distinguish the sounds, place the fork on your own head, and you may thus determine the difference.

ANATOMY AND APPEARANCES OF THE HEALTHY MEMBRANA TYMPANI, IN THE LIVING.

Having considered at some length the best available means of exploring the parts of the hearing apparatus accessible to inspection and instrumental tests, we will once more address our attention to the membrana tympani, and the appearances which this most important part presents when brought clearly into view.

The anatomist, who obtains his knowledge of this structure from *post-mortem* examinations or from dried preparations only, can have no very clear idea of its condition during life; for its delicacy, its fine epidermic covering, transparency, curvature, colour, and lustre, are all more or less destroyed or altered in the dead. The descriptions of the membrane which you might meet with in anatomical text-books are, on this account, mostly untrustworthy and insufficient for the requirements of the aural surgeon.

The membrana tympani may be described as a thin elastic membrane, stretched obliquely across the further end of the external auditory canal in such a way as to form an acute angle with the lower wall, and an obtuse angle with the upper wall, of the latter. It is inserted into an osseous groove, which is not quite complete above. We have to notice, separately, its *relative position, inclination, form, size, colour, curvature, lustre, degree of transparency,* and *texture;* for in all these particulars are variations perceptible in disease. My description will be as brief and condensed as possible.

As regards *relative position* of the membrane. If examined on the outer side, we first remark the handle of the malleus, extending as a whitish yellow stripe from the upper border downwards and backwards very nearly as far as the centre, where it somewhat expands and is flattened. This *manubrium* (erroneously called by Toynbee and others the long process), inclines inwards towards the cavity of the tympanum, and divides the membrane into two rather unequal parts, the anterior and posterior, of which the first is slightly smaller than the other. We may assist our idea of the inclination of the manubrium of the malleus, by dividing the membrane into four quadrants (*vide* engraving page 43).

In the anterior upper fourth at the centre of the periphery we see the very projecting, small white rounded tubercle, the short process (*pro-*

cessus brevis), to which the neck of the malleus and the tensor tympani muscle are attached on the inside; then the handle, curving downwards towards the centre between the fibrous layers of the membrane, draws it inwards and arches it in such a manner that it presents a concavity, and thus is determined what is called the curvature of the membrane. This curvature is greatest towards the lower end of the handle, and is termed the *umbo* or umbilical contraction. Although the membrane, as a whole, is thus made concave externally by the traction of the malleus handle, we shall upon closer examination perceive that it nevertheless exhibits a tolerably well defined convexity in some parts, especially at the anterior and lower portion. This is somewhat difficult to understand, and still more so to describe; but this peculiar arrangement will account for the many variations in the form of the little *cone of light*, or triangular reflection, which is visible in certain movements of the membrane during air-pressure alterations, in inflation or catheterisation from within, and in disease.* The short process of the malleus produces at the upper-anterior quadrant a decided bulging outwards, from which generally proceed two folds, the shorter running forwards and the longer backwards.

The *inclination* of the membrane is also difficult to understand. In the fœtus and young children, the

* This interesting appearance could not be represented in the accompanying woodcut, which was made from a photograph of the parts removed and separated after death. The reflected bright spot is therefore, not seen.

Fig. 1.

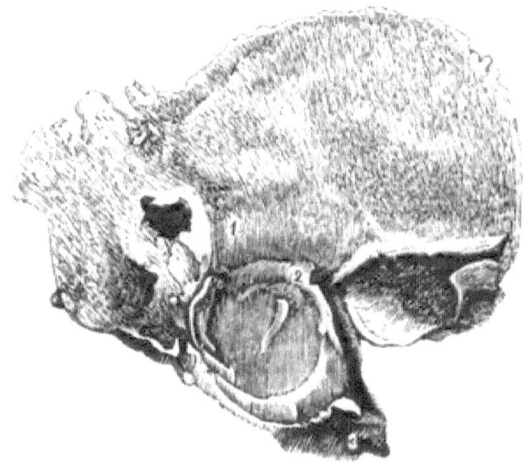

THE MEMBRANA TYMPANI, ETC.

(Seen from the outside.)

1. The right temporal bone of the infant. 2. The membrana tympani, showing just below (2) the short process (*processus brevis*) of the malleus and its handle (*manubrium*) descending to the centre of the membrane. 3. The *petrous portion* of the temporal bone which contains the internal ear, or labyrinth.

tympanic ring into which the membrane is inserted like a watch-glass into its case, lies nearly horizontal with the meatus, and though it gradually approaches a more vertical direction in the adult, yet the amount of its inclination varies much in different individuals. Our judgment of its size or superficial area is modified by this inclined position. If it be very oblique to our axis of vision, it appears to us smaller than in reality, being of course seen in perspective, or foreshortened; and the less the inclination, the larger does the membrana tympani appear to us. The inclination is relative to the walls of the external meatus, and the measurements of the angles thus formed have been variously stated, but I think it may be estimated that the membrane forms, with the upper wall, an obtuse angle of about 140, and with the lower, of course an acute angle of 40 degrees.

The *form* of the drum-head varies with the diameters of the osseous meatus, and in different individuals. In the child it is nearly circular, in adults more oval, sometimes irregularly heart-shaped, owing to the interruption in the surrounding osseous ring.

The *size* likewise varies in accordance with the form and size of the osseous ring. In the fœtus, as you here see, the size of the membrana tympani bears no proportion whatever to the body, since it has nearly completed its growth at birth, and mostly afterwards only increases in length. In the adult its average vertical diameter is between 9 and 10 millimetres, and its horizontal diameter about 8 or 9.

The *colour* of the membrane must in truth be

studied on the living subject, as maceration after death alters it considerably. Hence the discrepancies between the accounts in various text-books. Another source of disagreement lies in the differing methods of investigation adopted by authors: the age of the subject, the kind of light used in the examination, the degree of transparency of the membrane, and its inclination to the auditory canal, are all elements which go to make up the colour, and greatly influence it. Nor can we, properly, speak of the membrane as a whole, since the different parts of it are, with the same light, differently tinted. In what may be characterised as a normal condition of the membrane, the colour viewed in ordinary daylight, most nearly approaches a *neutral grey*, mingled with a deeper tint of violet and light yellowish-brown. The grey is darker at the anterior portion than at the posterior. When the membrane is transparent it has a yellowish-grey colour, near its centre, at the end of the manubrium of the malleus, and a little behind it, on account of the yellow rays reflected from the promontory. In some very rare cases of transparency and thinness, the long process of the incus and even the posterior crus of the stapes, may be distinctly seen. In children, the dermoid layer on the outside, and the mucous membrane on the inside, are both of them thicker than in adults; and hence the membrane appears of a darker grey and duller, and the promontory is rarely seen through it. In old age it is less translucent, whiter, and lustreless.

Curvature and Lustre.—Differences in inclination

and curvature affect in many ways the lustre of the membrane. We have noticed its peculiar form, and that as a whole it is concave externally; but from the deepest point of its concavity (the umbo) towards the circumference, it curves somewhat forwards, and is slightly convex. Now, in healthy membranes, in addition to the general lustre extending over the whole surface, there is always to be seen a remarkable spot of bright light upon the anterior inferior quadrant. This "speck of light" (the "Lichtkegel," or "cone of light," of the Germans) was first described by Sir William Wilde. It is triangular in form, having its apex at the end of the malleus handle, and its base towards the lower and fore part of the membrane, near its periphery.

A knowledge of all the variations in the position of this cone of light is of the utmost importance, inasmuch as its form and size not only indicate alterations in the membrane itself, but also the state of the Eustachian tube, whether freely pervious to air or not. You constantly perceive that I draw your attention to the alterations of this spot when examining a patient. For instance, while the attempt is being made to force air into the tympanum by the Valsalva experiment, or when during the act of swallowing with nostrils closed the internal pressure is lessened, we become assured that the tube is open, if we clearly see the cone of light change its form. It will be found that in the greater number of cases this triangular reflection is broader at the base when the drum-head is pressed outwards, and smaller when it is drawn inwards by rarefaction of the tympanic air. Sometimes you see it as a circular spot in other parts of the membrane; sometimes divided, or striped; and occasionally it is altogether absent. As surely as there is any abnormal change in the curvature or lustre of the membrana tympani, so surely does this little triangular spot vary in form; and so far, its appearance is of great diagnostic value.

As the *lustre* of the outer surface of the membrana tympani depends upon the healthy state of the epidermic layer, there is a general loss of it when the delicate transparent cells of the epidermis are thickened, macerated, or infiltrated. The cone of light, both in the living and dead subjects, may therefore be compared in its appearance to the reflected light from the cornea. Different views prevail among authors respecting the cause of this light spot; but Politzer has, I think, correctly explained it. He says:—" The chief cause is the inclination of the membrane to the axis of the auditory canal, together with the concavity of the membrane produced by the action of the malleus." He carefully prepared a healthy tympanic membrane with the auditory canal removed, leaving it attached only to the bony ring around it. He turned it about so that other portions of the drum-head were successively brought into the position of the original cone of light; and he then found that every one of these points showed a reflection of light which varied only in degree from the usual form of light spot. You may convince yourselves of the correctness of Politzer's opinion by stretching any glistering animal membrane over a large ring, and giving it the inclination of the membrana tympani. If you then hold the mirror relatively as you do in examining the drum-head, you will perceive no reflection; but if the central portion be drawn in by pressure or traction, the reflection will be immediately seen at the exact spot corresponding to where the cone of light is formed on the membrana tympani.

Transparency.—From the absence of the usual transparency arising from opacities of the membrane, we can scarcely draw conclusions as to the nature of the lesions which may have produced the change. Sometimes it may have been caused by morbid processes on the external surface which have run their course, and resulted in complete recovery; in other cases, a catarrh of the tympanic cavity has passed off, leaving the mucous layer opaque, but has not impaired its function.

Texture.—Turning now to the histology of the membrana tympani, we find it to consist of three layers,—a middle fibrous (*lamina propria membranæ tympani*), an external dermoid, and an internal mucous layer; these latter being the two coverings which it receives from the meatus and from the cavity of the tympanum respectively. The *dermoid* layer is an extremely delicate continuation of the lining of the meatus, and consists of very transparent epidermis, and the elements of the cutis in which the blood-vessels and nerves ramify. In inflammation these vessels give the red appearance, and the exquisite sensibility or extreme pain when inflamed depends upon the supply of nerves to the same layer.

It was formerly thought that the fibrous layer secreted the epidermis, and the existence of this cuticular layer now accounts for various phenomena in certain diseases of the ear which the surgeon could not previously understand. The proper middle *fibrous* layer can be easily separated into two laminæ, which are named from the arrangement of their component fibres, an outer *radiating* and an inner *circular* set.

The outer radiating fibres run from the cartilaginous ring which is inserted into the osseous groove of the temporal bone (before spoken of), to the handle of the malleus. They slightly increase the thickness of the membrane towards the centre, because here their fibres are more closely packed together; while the thinner are on the posterior part of the membrane between the circumference and the handle. The inner or circular layer of fibres are arranged, as their name implies, in concentric circles. They are very strong and firm at the circumference, but at the centre so attenuated as to be scarcely perceptible. This layer is very intimately connected with the mucous membrane, and is nourished by it. The handle of the malleus lies between the radiate and circular fibres, and is inserted (as it were) through a slit in the latter; so that the uppermost rings of them are anterior and external to the neck of the malleus; while the lower and larger portion of them are posterior and internal, thus surrounding the neck like a ruffle. Grüber states (and for reasons to be assigned later, I think correctly), that there is a complete articulation between the cartilaginous structure at the circumference of the membrane, and the malleus; that they are separated by an epithelial layer; and that there is synovial fluid between them.

The *mucous* layer is a continuation of the mucous membrane of the tympanum, and, in a healthy state, is extremely thin, consisting almost entirely of a single layer of pavement epithelium; but it is very frequently indeed subject, in aural catarrh, to pathological changes, which soon thicken it. There are plenty of blood-vessels in the mucous membrane, but few or no recognisable nerves. It differs greatly in this respect from the external cutaneous layer derived from the meatus; and this fact coincides with our practical experience before referred to, that superficial and external inflammations of the membrana tympani are always painful, while the greatest changes may occur in the internal or mucous layer without the patient's suffering any pain whatever in the ear. You cannot over estimate this circumstance.

THE BLOOD-VESSELS OF THE MEMBRANA TYMPANI.
—There are two sets of vessels derived from two distinct sources, and they are completely separated from each other, except at the circumference of the non-vascular fibrous structure (the *lamina propria*) where they freely anastomose.

The external vascular net-work of the drum-head belongs to the outer cuticular layer, whilst the inner belongs to the mucous membrane of the tympanum (the mucous layer). The external plexus is derived from the deep auricular branch of the external carotid (the *arteria auricularis profunda*). These vessels extend themselves from the upper wall of the auditory canal upon the membrana tympani, and radiate upon its periphery, so as to form a complete wreath by uniting there with the vessels from the tympanic cavity. Some of the branches run down the centre of the membrane, and oftentimes are visibly filled with blood when we syringe the ear with warm water, or when the patient inflates.

The inner vascular net-work arises from the vessels of the tympanic cavity. It is smaller, and is more of a capillary system than the outer. The plexus is derived from the tympanic branch of the stylo-mastoid, and freely anastomoses with the deep auricular artery (which comes from the internal maxillary) on the circumference of the membrane, particularly at the upper deficiency of the tympanic ring, forming with these external branches the circular plexus which gives the vivid red appearance to the membrane in acute inflammation.

One or two arteries run along the handle of the malleus, and deserve especial notice. Occasionally, when I am in doubt as to the position of the malleus, owing to opacity, &c., of the membrane, I am guided to its handle by the red lines or lines coursing downwards on it or parallel with it. You will scarcely ever fail in making this manifest by directing the patient to attempt to inflate the drum; for whether he succeeds or not, the straining will cause turgescence of the vessels of the membrane, and trace out the course of the manubrium.

THE NERVES OF THE MEMBRANA TYMPANI.— The nerves are chiefly, if not wholly, to be found in the cuticular layer. The tympanic nerve is a branch of the superficial temporal or auriculotemporal, which comes from the sensory part of the third branch of the fifth; and it is this connexion which causes the great sensitiveness of the outer surface of the membrane. Nerves have not satisfactorily been made out to ramify either in the fibrous or mucous layers, but the external cuticular layer of the membrane has numerous small branches derived from the same important inferior maxillary, or third division of the *trigeminus*, as that which supplies the tensor tympani. This common origin of the nervous supply to the muscle which moves the membrane, and to the surface of the membrane itself, is very suggestive. I think it supports the view that the reflex action of the muscle under the stimulus of certain sounds, as explained in my last lecture (page 34), is due to the nervous power imparted both to muscle and membrane, by this compound nerve.

LECTURE IV.

THE CAVITY OF THE TYMPANUM.

GENTLEMEN,—I have more than once remarked in these lectures, that the condition of the membrana tympani will, as a rule, furnish not only most valuable evidence of the nature of any disease going on in its own proper structure, whether on its external or internal surface, but also afford visible signs of any catarrhal affections existing in the middle ear.

It can scarcely happen that an aural catarrh should locate itself in the cavity of the tympanum, and damage the functions of the ossicula and fenestræ of the labyrinth to any considerable extent, without being recognisable by inspection of the membrana tympani. I have therefore thought it undesirable to classify inflammations, whether acute or chronic, according to the old nosological arrangement, into otitis interna and externa: these terms applying to inflammatory diseases on the outside and inside respectively of the membrana tympani. Such a division will neither assist our diagnosis nor improve our treatment. The nomenclature adopted by aural surgeons may be very advantageously simplified; and on this account, in considering the pathology of aural catarrh, and the most common of all affections of the ear, I shall describe the symptoms, differential diagnosis, results, and treatment, simply according to the situation of the parts affected, whether they be the cavity of the tympanum or its external membrane, the mastoid cells, or Eustachian tube. For, if we critically examine the lengthy cases of inflammation of the membrana tympani detailed by surgeons, it will be seen that this structure has not been primarily and independently attacked, but that a diffuse dermoid inflammation of the auditory canal, or catarrh of the tympanum, has extended itself by continuity of surface on to the drum-head, and only secondarily involved it in disease. As, therefore, we cannot separate the pathology of the middle ear from that of its external boundary wall (the membrana tympani) I think it well that a general sketch of the anatomical relations of the drum to the adjacent parts, should follow our remarks on the appearances and structure of the drum-head.

If the membrana tympani were not stretched across the bony part of the auditory passage, the drum and the meatus would form one cavity. Such is the case presented to us in many diseases where ulceration and suppuration, the products of inflammatory action, have partially or wholly destroyed the delicate tympanic membrane. The space included by the cavity of the drum is that between the membrana tympani and the labyrinth, or internal ear. It is filled with air, and has the three ossicula auditûs (the malleus, incus, and stapes), which are firmly articulated with each other, extending across it (Fig. 2). These form an elastic vibratile chain, connecting the outer with the inner ear. The first of these bones, the *malleus* (Fig. 2, No. 6), has its handle (or manubrium) embedded in the fibrous layers of the membrana tympani, while its head articulates with a corresponding depression in the body of the second bone, the *incus*. This second bone (Fig 2, No. 4) has two processes; one horizontal, which forms a kind of universal joint at its extremity, where it rests in a little hollow in the roof of the tympanum, near the mastoid cells; and the other, vertical and longer than the horizontal, descends parallel with the handle of the malleus (but not quite so far downwards) to articulate with the rounded head of the third bone (Figs. 2 and 3, No. 8), the stapes. This last bears, perhaps, a more accurate resemblance to the stirrup, from which it derives its name, than any other part of the body which has been fancifully likened to some familiar object. Its base or foot is inserted into the fenestra ovalis, where it is only separated from the labyrinth by the most thin and delicate periosteum of the vestibule. Text-books inform you that this bone is inserted into a membrane like the external membrana tympani, and term the latter the "membrana fenestræ ovalis." This statement requires modify-

To face page 54.

FIG. 2.

THE RIGHT TEMPORAL BONE OF THE INFANT, AND THE MEMBRANA TYMPANI.

(Seen from the inside, in connection with the ossicles [natural size]. The pars petrosa has been removed.)

4. The incus. 5. The long process of the incus, articulating with the stapes (8). 6. The malleous, articulating with the incus (4). 7. The *processus gracilis* of the malleous. (7) is on the tympanic ring, into which the membrana tympani is inserted. 8. The stapes. The base, or foot-plate, of the stirrup, is from this point of view most prominently seen.

FIG. 3.

THE OSSICULA AUDITUS ISOLATED.

1. Malleous.   2. Incus.   3. Stapes.

ing. Toynbee described a complete joint between the stapes and the fenestra ovalis, "stapedio vestibular articulation," with all the parts constituting a joint, such as cartilage, ligament, and synovial fluid. I do not quite coincide in this opinion, for in almost all the specimens I have examined, a thin plate of bone projects from the inner end of the oval fenestra (or window), and forms as it were a ledge, on which the margin of the stirrup foot may rest, if too suddenly driven inwards by violent vibrations of the ossicular chain, influenced by corresponding movements of the membrana tympani.

It is evident that all these three bones must serve to convey sound-vibrations from the drum-head to the fluid in the labyrinth at the fenestra ovalis, which fluid will, finally, excite the fibres of the auditory nerve therein expanded. In order to enable the membranes at either end of this chain to be tightened or relaxed, its mechanism is controlled by two muscles, the action of which has been previously described. One, the tensor tympani, is affixed to the malleus (Figs. 4 and 5, Nos. 1 and 1), and is capable of pulling its handle inwards, so as to render the membrana tympani tense, and thus moderate the effect of very loud sounds, in the same manner that the circular fibres of the iris modify the intensity of light to the eye. This muscle, I believe, is not generally under the control of the will; its contraction being stimulated by nervous filaments from the otic ganglion, which are derived from the motor root of the inferior maxillary of the fifth. When, by the action, indirectly, of this **muscle**, the base of the stapes is pressed inwards, towards the vestibule at the fenestra ovalis, it is manifest that the fluid of the labyrinth must be made to move, and as the only yielding point is at the other end of its bony cavity at the fenestra rotunda, the membrane there situated must bulge outwards.

This membrane of the *fenestra rotunda*, called also the *membrana secondaria*, then becomes tense, like the membrana tympani, and in a manner similar to it is placed in a condition either to receive and transmit sounds of a high pitch of tone, or, if tightened sufficiently, will prevent vibrations of too excursive a character. Thus, not only are the sounds impinging upon the membrana tympani controlled, and prohibited from injuriously affecting the vestibular nerves by way of the fenestra ovalis, but those vibrations which are contained in the cavity of the tympanum are prevented from too violently affecting the more delicately organized cochlear nerves, by the way of the fenestra rotunda.

The other muscle (*see* Fig. 5, No. 4), the stapedius, is in a manner antagonistic to the tensor tympani, although in certain instances their action is combined. The functions of these muscles, however, would require more time to explain thoroughly than can possibly here be bestowed. I may just note that I have for many years entertained an opinion that the stapedius muscle was capable of relaxing one part, at the same time as it tightened another part of the membrana tympani; and the discovery by Grüber of a little joint between the malleus-handle and the drum-head seems to confirm this

Fig. 4.

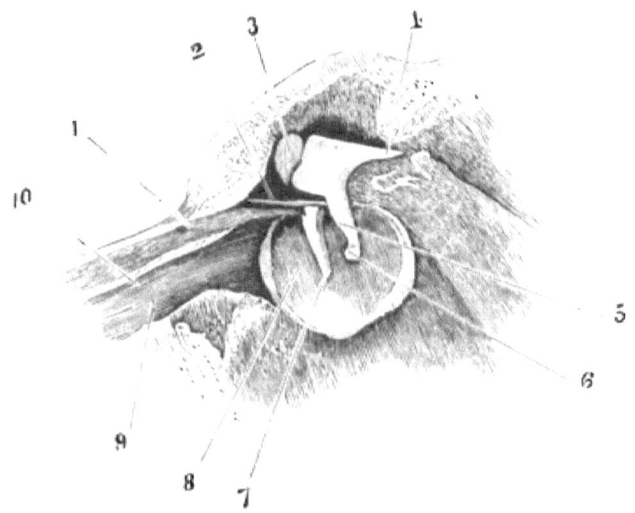

EXTERNAL WALL OF THE CAVITY OF THE TYMPANUM (RIGHT SIDE).

Seen from its internal face. The Eustachian tube and canal for the tensor tympani muscle are laid open.

1. Tensor tympani, with its tendon inserted into the manubrium of the malleous. 2. Chorda tympani nerve. 3. Head of the malleous. 4 and 5. Short and long processes of the incus. 6. Lenticular process of the incus. 7. Extremity of the manubrium of the malleous. 8. Membrana tympani. 9. Eustachian tube. 10. Upper wall of Eustachian tube. (After Henle.)

Fig. 5.

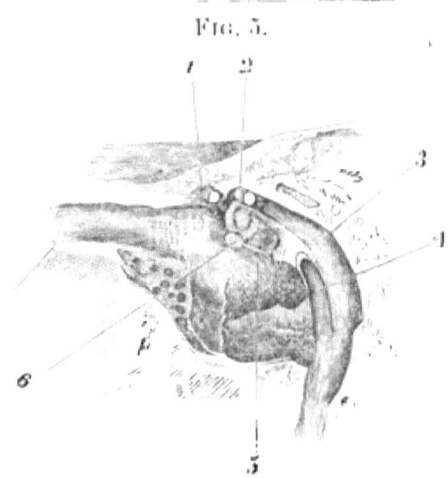

THE INTERNAL WALL OF THE CAVITY OF THE TYMPANUM (LEFT SIDE), WITH THE STAPES.

The aqueduct of Fallopius and the canal for the stapedius muscle are laid open.

1. Tendon of the tensor tympani cut close to its entrance into the tympanum. 2. Facial nerve, cut. 3. Aqueduct of Fallopius, open. 4. Stapedius muscle, within the pyramid. 5. Promontory. 6. Head of the stapes. 7. Eustachian tube.

idea. By this means, every undulation of sound which the auditory nerve is capable of perceiving, can be propagated to it. The stapedius muscle is supplied by the facial nerve (*see* Fig. 5, No. 2, page 56), and, of course, can be put into voluntary action.

It must never be forgotten that the two fenestræ are the most important portions of the wall of the drum, and that their integrity, as well as a healthy condition of the mucous membrane lining them, is essential to perfect audition. Any morbid change, especially thickening of the mucous membrane from catarrh of the middle ear, will lessen the mobility of the stapes in the oval opening, and the elasticity and vibratility of the still finer and more delicate membrane of the round opening, will likewise be limited or destroyed from the same cause. Unfortunately, both these parts of the auditory apparatus which conduct vibrations to the nervous or sound-perceiving portion of the organ, are very readily affected in common catarrh; and it is an abnormal condition in them, which furnishes the severest forms of deafness in patients who cannot hear the vibrating tuning fork or watch when placed upon the cranial bones. Pressure upon the labyrinth-fluid is the result of a confirmed or unrelieved catarrhal inflammation. For further description of the functions of the various component parts of the tympanic cavity, I would refer you to some articles of mine in the *Lancet* for the earlier months of 1869, but the subjoined diagram, taken from that publication, may assist in elucidating the subject.

Fig. 6.

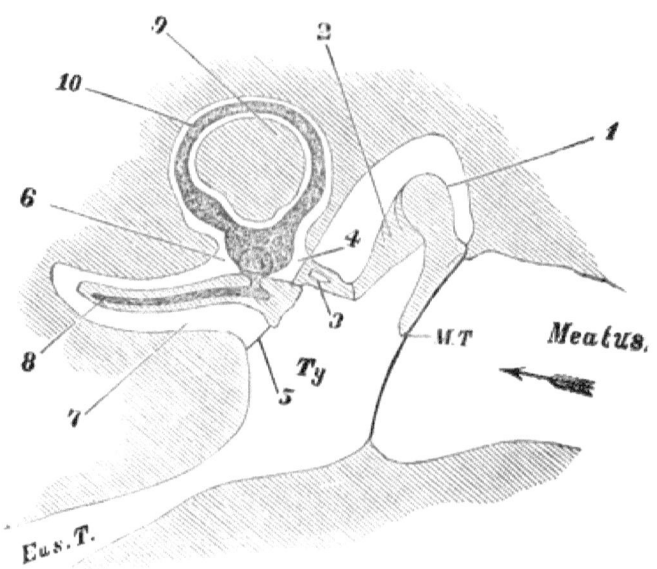

A DIAGRAM ILLUSTRATING THE RELATIVE POSITIONS OF THE VARIOUS PARTS OF THE EAR.

M. T., Membrana tympani. *Ty.*, Tympanum. *Eus. T.*, Eustachian Tube. 1. Malleus. 2. Incus. 3. Stapes. 4. Fenestra ovalis. 5. Fenestra rotunda. 6. Scala vestibuli of the cochlea. 7. Scala Tympani of cochlea. 8. Scala media (*canalis cochlearis*) of the cochlea. (The cochlea is supposed to be unrolled.) 9. One of the vertical semicircular canals and its ampulla between the figures 6 and 10. 10. Membranous labyrinth. Only one of the three canals is represented.

We next consider the other walls of the tympanum. The roof (*tegmen tympani*) is often thin and transparent, has lying upon its upper surface the dura mater, and therefore forms the partition wall between the tympanic cavity and the brain. Not only is there this contiguity of structure between the eardrum and cerebral lobes, but a free vascular communication exists between the vessels of the dura mater and the mucous membrane of the tympanum at this point. As the result of a catarrhal inflammation in the tympanum, disease may be propagated to the brain, either by continuity of structure, as in caries of the bone, ulceration of the dura mater, and ultimately abscess in the substance of the cerebrum, or it may extend to the membranes and sinuses through the vessels of nutrition, and thus form a collection of pus in the cerebral mass.

The *lower* wall of the tympanic cavity is in close vicinity to the jugular vein, which lies immediately underneath it in the jugular fossa, and disease of this part of the bone may render it so carious as to allow the coats of the vein to be in direct contact with the mucous membrane of the tympanum. In such a case the inflammatory process will have ready communication with the venous system of the cranium, and purulent infiltration may be the consequence.

One part of the *inner* wall of the tympanum, directly opposite to the membrane, may occasionally be seen, especially when the latter is transparent, or very much drawn inwards. This is the *promontory* (*see* Fig. 5, No. 5, page 56), a smooth broadish prominence, advancing into the cavity of the tympanum, and dividing the two fenestræ. This projection is caused by the first turn of the cochlea, and its surface is grooved for the tympanic branches of the glosso-pharyngeal nerve. In a young man now attending at the hospital, you have had an opportunity of plainly seeing that the membrana tympani in both ears is adhering to the promontory—a state of things not very uncommon as the result of neglected catarrhal inflammation.

Above and behind the fenestra ovalis is the *aqueduct of Fallopius* (see Fig. 5, No. 3, page 56) covered in by a thin plate of bone, which is sometimes porous and defective in parts. In this canal lies the facial nerve, or *portio dura* (see Fig. 5, No. 2), and thus is accounted for the facial paralysis occurring in caries, and in certain inflammatory or hyperæmic conditions of the mucous membrane.

Between the fenestra rotunda and the Fallopian canal just referred to, is the little conical eminence called the pyramid, from the apex of which there is a small opening for the tendon of the stapedius muscle. (*See* Fig. 5, No. 4).

On the *posterior* wall, close to the roof, are the openings into the mastoid cells, and it is very necessary to bear these anatomical relations in mind, when inflammation attacks the cavity of the tympanum. These cells have also a ready communication with the lateral sinus, by several small apertures through which veins pass to the sulcus lateralis, and small arteries likewise freely enter the dura mater which adheres internally to the superior surface of the mastoid process. It is important to remark that in childhood the mastoid process is in a rudimentary state, the only part of it represented being the horizontal portion adjacent to the tympanic cavity; and it is this particular part which so frequently becomes affected in early life, when disease invades the tympanum. The walls of both, are contiguous at their upper part, and as the dura mater and cerebral cavity lie immediately above, it is obvious that irritation of the brain or cerebral abscess may readily ensue from disease in the mastoid cells consequent upon scarlet fever, measles, scrofula, or even common catarrhal inflammation.

The MUCOUS MEMBRANE lining the cavity of the tympanum throughout is very thin, delicate, and smooth. It is spread over every surface, is reflected round the little bones, ligaments, tendons of the muscles and nerves, it covers the internal surface of the membrana tympani, and the membrane of the fenestra rotunda. It extends into the mastoid cells, and is continuous through the Eustachian tube, with the great gastro-pulmonary mucous membrane; possessing thus an extensive and intimate superficial connection with the respiratory and digestive organs. We can well understand, therefore, how in derangements of almost any, even remote parts of the internal surfaces, morbid impressions may be conveyed either by sympathy or continuity of tissue, to the centre of the ear, and there develop symptoms which may considerably impair the function of hearing. It is in this manner that bronchitis, catarrh, cutaneous affections, or dyspepsia, may (by their ill effects upon the mucous surfaces) become the cause of deafness.

The mucous membrane itself is extremely delicate, and has minute epithelial cells on its surface, some of which are ciliated. In fineness and smoothness of texture it approaches the characters of a serous membrane, and from its tenuity, is so transparent, that the nervous filaments upon the promontory are distinctly seen, as well as the crura of the stapes and their connection with the base. In children the membrane is highly vascular, but with adults its colour in a healthy state, is a whitish-gray, and polished. Like all mucous tissue, it secretes mucus, but in so small a quantity, and so thin, that it is scarcely perceptible, and so can find a ready outlet through the Eustachian tube. It is necessary to know that the condition of the tympanum is very different in the fœtus and for some time after birth. The cavity is filled, not with air, but with a "cushion of mucous tissue" (Virchow), which swells out from the labyrinth wall to the membrana tympani upon the inner surface of which it rests. This anatomical fact, as well as almost daily practical experience, proves how favourable to the development of inflammatory affections is the condition of the middle ear in children. Ear-ache is of such common occurrence that scarcely a child escapes it at one time or another, and examinations have abundantly shown that it depends upon catarrhal inflammation of the tympanum, or of the adjacent surfaces of the membrana tympani and external meatus. We now term this disorder "Infantile aural catarrh."

THE EUSTACHIAN TUBE.

The Eustachian tube (Figs. 4, 5, and 6) extends from the anterior portion of the tympanum, exactly opposite the mastoid cells, in a direction downwards, forwards, and inwards. It should be borne in mind that this tube does not run from the very floor of the cavity, but from about half the way up, so that if fluids be injected through it with sufficient force, they will pass into the mastoid cells, and also partially remain at the bottom of the tympanum. For this reason I object to the frequent injection of medicated or other fluids. The tube consists of a bony and a fibro-cartilaginous portion, and in this respect resembles the external auditory canal; but the proportionate length of the two parts is here reversed. In the Eustachian tube, the length of the osseous part is from half to three-quarters of an inch, while the membranous is about an inch. Its faucial orifice is the widest and most dilatable; but further up, where the cartilaginous and bony parts unite, it is so small as scarcely to admit a probe. This point is often called the isthmus of the tube. The tube itself is thus trumpet-shaped, having a gaping orifice, which usually is situated a trifle higher than the floor of the nares; but it occasionally varies in position, and when the naso-pharyngeal membrane is tumefied, or the inferior turbinated bone projects a little over the orifice, difficulty is experienced in guiding instruments into it.

The mucous membrane lining the Eustachian tube varies in structure. At the faucial end it is very

4*

thick and puffy, and sometimes lies in such folds as to obstruct the free entrance of air into the tympanic cavity, either when attempted to be forced in by the catheter or by other methods hereafter to be referred to. On account of these folds and of some largish glands, you often see me feeling, as it were, my way into the opening at the pharynx; but seldom will the surgeon, practised in catheterism of this part, fail to insert the point of the instrument, if he do not employ force or haste in his manœuvres. As the tube extends upwards towards the tympanum, the lining mucous membrane becomes thinner until it reaches the tympanic orifice, where also there are glands, and it is more vascular. Further (and the best) description of the tissues composing the tube and their microscopical appearances may be found in Dr. Von Tröltsch's work "On the Diseases of the Ear."

The chief use of the Eustachian tube is to provide ingress and egress of air to and from the tympanum, so as to bring about a regular interchange of air between that contained in the naso-pharyngeal cavities and that within the drum. It is now generally believed by anatomists that the surfaces of the Eustachian tube usually lie in contact, and that the too ready ingress of atmospheric air to the tympanic cavity is prevented by the tensor and levator palati muscles; also that the principal function of these muscles is to open the tube during their contraction in the action of swallowing. It necessarily follows that whenever this act is performed, the Eustachian tube must be opened. Afterwards, its lips again fall together, and no air can enter or recede from the drum. It is obvious that the tympanum must thus be generally a closed cavity, and will contain air of the same density as that on the outside, air only entering and escaping from its interior in such proportion as is needful for maintaining the same tension on both the inner and outer side of the membrana tympani. Without such an arrangement for the renewal and equalisation of air, the requisite vibratility of the tympanic membrane could not be kept up,—the air would become exhausted or absorbed, the membrana tympani and ossicles would fall inwards, and cause pressure upon the vestibular fenestra and labyrinth-fluid. You frequently see this state of things as the consequence of catarrh affecting the mucous lining of the Eustachian tube, and causing an obstruction to the passage of air into the drum; and you also often have the pleasure of witnessing the restoration of the lost hearing by the use of a simple and efficacious contrivance,—my modified Politzer bag, which will be again referred to.

I have in another place shown that during the act of swallowing is not the only time at which the tube-muscles contract, and allow the entry or recession of air. We can all, doubtless, with a little practice, gain such command over the palate muscles as to be able to open the orifices of the tubes. Yawning, violent blowing of the nose, and some peculiar voluntary motions of the palatine arches during a forced inspiration or expiration (habits which miners acquire in descending shafts), will open the tubes, and allow air of the same character and density as that which is pressing on the external membrana tympani, to enter the drum and equalize the pressure upon both its surfaces.

There is another way of inflating the tympanum through the Eustachian tubes, which you constantly observe me directing patients to attempt during examination in a supposed case of impermeable tube. It is called the Valsalva method, after its discoverer, and consists in an effort at pressing air into the drum by a forcible expiration through the nostrils while they are held closed. If the tubes be permeable, air is thus forced through them into the drum, in such excessive quantity and with such power as to overcome the muscles of the tube, and the elastic resistance of the membrana tympani. The latter will thus be blown outwards. If, during this act, we place the otoscope in the external auditory canal, we recognise the peculiar snapping or crackling noise of the stretched membrane, and judge thereby of the degree of freedom or difficulty with which the air enters from the inside. As soon as the air has entered the drum, and force is no longer applied by the patient, the tube muscles return to rest, and the air becomes imprisoned in the tympanic cavity. Besides the visible movement of the membrana tympani occurring while this inflation is going on, and the noise produced by its being suddenly stretched, we may in another way judge of the permeability of the Eustachian tubes, by using a manometer. Here is the instrument which I modified and got made after the suggestion by Politzer, and I have found it very useful in determining the exact states of the membrana tympani when an inspection could not be obtained.

FIG. 7.

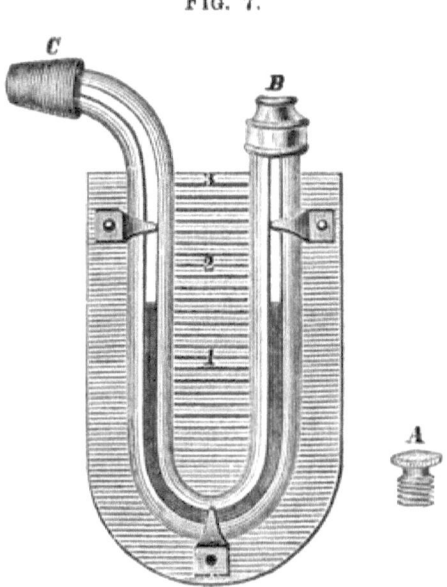

TYMPANO-MANOMETER (ALLEN'S).

A screws down tight upon *B* to prevent disturbance of the coloured fluid when the Manometer is carried in the pocket. *C* represents a piece of india-rubber tubing to fit air-tight in the Meatus. The instrument can be reversed on its graduated boxwood scale, so as to be applied to either ear. (Made by Weiss and Son.)

If it is placed air-tight in the meatus, and a forcible attempt at expiration is made with the nose closed, the coloured fluid in the manometer will be observed to rise, if the Eustachian tubes are patent; in other words, a *condensation* or increase in the volume of the tympanic air will be proved to have occurred.

Now, if by another experiment, presently to be described, the tympanic air has been drawn out through the Eustachian tubes, a *rarefaction*, or diminution thereof will be proved by a fall of the fluid in the manometer. Mr. Toynbee asserted that during the simple act of swallowing with the nostrils closed, the air in the drum is condensed (*i.e.* increased), because we experience a sense of fulness in the ears, and a peculiar cracking noise is heard at the same time.* This erroneous supposition unfortunately led him to discard catheterism as a means of diagnosing the permeability of the Eustachian tubes. We now know that if we swallow with the nostrils closed, a state the opposite to that which he imagined, really follows. The air is rarefied, or lessened in quantity, and this is made evident by the manometer, the fluid in which slightly falls after this act is performed. In the cases of very thin membranes and cicatrices which have fallen inwards towards the promontory of the tympanum, we may sometimes perceive that there is a momentary bulging out of these depressed parts, but immediately afterwards they sink further inwards. The fluid in the manometer rises quickly and considerably, but then falls slightly, resting a little lower than where it previously stood.

I have been desirous to particularise these movements of the instrument, because my own deductions from them differ to some extent from the views of other investigators; and I am also anxious to show

* This sound is due to the action of the palate muscles, not to the membrana tympani.

you that my modified tympano-manometer will really indicate the variations in the pressure of air within the tympanic cavity, as well as the patency or otherwise of the Eustachian tube.

In the next lecture I shall describe the new and very important way of inflating the ear, which is called after its inventor, the Politzer method; the greatest discovery made of late years. Also, how catheterism of the Eustachian tube is performed.

LECTURE V.

ON THE DIFFERENT METHODS OF EXAMINING AND INFLATING THE MIDDLE EAR.

GENTLEMEN,—It would be inconvenient and confusing were I to defer description of the various means employed to inflate the middle ear, or overcome obstructions in the Eustachian tube, until the details of treatment should render it necessary for us to particularise one or other of them. I therefore purpose now to give you a summary of the different methods in use for examining, diagnosing, and likewise removing certain conditions of the tympanum which occur in aural catarrh. Subsequently, in referring to the treatment to be adopted in each case under consideration, it will be only necessary for me to designate the *mode* intended, by mentioning the name of its inventor. The results to be obtained are almost identical with those produced by catheterism of the Eustachian tube, but the choice of the substitutes for that procedure (which cannot in all instances be had recourse to), will be thus brought under your notice without repetition of details. The diagnostic value, as well as comparative efficiency of each method, will also thus be learnt in the beginning. You witness the application of all of them by turns on the patients here, as their cases appear to demand, but there is not at all times convenient opportunity to explain my reasons for employing one way in preference to another. The frequently happy result, especially in the cases of children and delicate women, which follows the use of the inflating-bag in the treatment of aural catarrh, is the best testimony to its efficiency and perfect applicableness.

I must first of all observe, that another instrument is used in diagnosis, to enable us to judge of the effect produced on the Eustachian tube, or in the middle ear and membrana tympani, when the drum is inflated by any of the methods presently to be specified.

When we desire to ascertain the sounds produced in any viscus or hollow part of the body by the functional movements of its contents, whether in the condition of health or disease, we AUSCULTATE it, either *immediately*, as by placing our ear on the patient's chest, when we listen for the pulmonary or cardiac sounds, or *mediately*, as by applying the stethoscope to map out certain portions. In like manner it is easy to auscultate the tympanic cavity when under experimental investigation by any of the different methods employed in inflating it. A stethoscope may be placed directly over the ear of the patient by another person, or the surgeon may avail himself of a more useful instrument, called by its inventor, Toynbee, the *Otoscope*, or diagnostic tube. This (differing from the otoscope described in a former lecture), consists of a single elastic tube, about eighteen inches or two feet long, and tipped at each end with ivory, ebony, or vulcanite. One

Fig. 8.

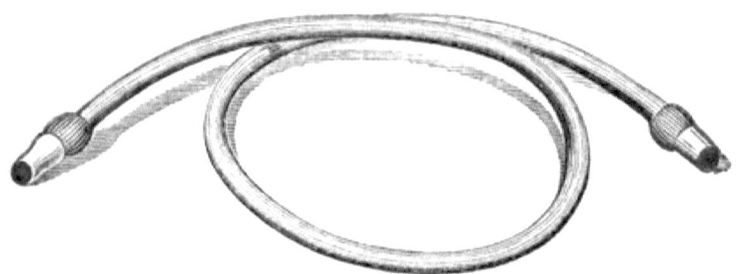

OTOSCOPE, OR DIAGNOSTIC TUBE.

end is inserted firmly into the patient's ear, and the other into that of the surgeon, who must be careful that no part of it touch his clothes or any of the neighbouring objects. Toynbee used it almost exclusively for the purpose of ascertaining the permeability or otherwise of the Eustachian tubes during an act of swallowing with the nostrils closed; and he imagined it would afford him the required information without resorting to the catheter.

The results obtainable by this process are meagre when it is compared with other diagnostic methods, but it has nevertheless its uses; for if the membrana tympani be viewed with the help of good reflected light while the patient swallows with his mouth and nose closed, we see the triangular "spot of light" alter its form if the Eustachian tube be permeable and the air in the tympanum has become somewhat rarefied (as previously described on page 68). Seeing any change in the *position* of the membrane, we are, of course, convinced that air has either entered or been withdrawn from the drum; and this fact, and some variability in the perceptible movements of the drumhead, is about the whole of what may be ascertained by Toynbee's method. He states that he perceived most distinctly "a faint crackling sound produced apparently by a slight movement of the membrana tympani . . . . . the entrance of air can be heard . . . . . the membrana tympani is seen to be pressed slightly outwards," &c. The *sounds* here spoken of may be heard during an ordinary act of swallowing with the nostrils left open, being produced simply by the contraction of the palate muscles, and pulling apart of the Eustachian tube walls. For the above and other reasons, before explained at length, I must advise you to reject Mr. Toynbee's theory of the *condensation* of air in the tympanic cavity during the simple act of deglutition with the nostrils closed. The diminution or *rarefaction* of air in the drum by Toynbee's process arises in the following manner: While the Eustachian tubes are patent, the air in the naso-pharyngeal cavity, being prevented from escaping through the nose, is pressed into the drum with sufficient force to overcome the resistance offered by the fibrous structure of the membrana tympani, and the latter is in consequence quickly pushed *outwards*. Before, however, the muscles which have opened the Eustachian tube (the tensor palati more especially) return to a state of rest, the drum-head, by its own inherent elasticity, aided by its own muscle (the tensor tympani) springs back to its position, even a little further inwards than before, and the result is a diminution of the contained air in the cavity. If the patient then make a swallowing movement with the nostrils *unclosed*, equilibrium will be restored between the air on the inside of the membrane and that on the outside, because a fresh supply of air will enter from the fauces as soon as the mouths of the Eustachian tube are opened. These results were made visible to you on the manometer, at my last lecture. Thus, the effects produced upon the membrana tympani, by the Toynbee method of inflation, differ wholly from those which follow upon the Valsalva process, and, in diagnostic value, Toynbee's method is the inferior.

The VALSALVA METHOD has already been partially explained on page 66. It is often practised, of their own accord, by persons hard of hearing. They find that a certain amount of improvement in their hearing takes place after pressing air into the drum and stretching the membrane, and they resort to it because of its therapeutical value. We, in examining patients, employ this method for the purpose of diagnosis. Its usefulness is, however, not limited to testing the permeability of the Eustachian tube (as we are taught by some aural surgeons who advocate catheterism too exclusively); but it also may inform or assure us as to the existence of a perforation in the membrana tympani, when we cannot by inspection detect one. It is a matter of everyday occurrence, that obstacles in the external meatus, fungoid growths, &c., covering the membrana tympani, prevent the eye from discerning an aperture in it, whereas by this simple experiment, when completely performed, air will be blown with a more or less whistling sound, through the Eustachian tube and cavity of the tympanum into the external auditory canal. If at this time the surgeon place one end of the otoscope in the patient's meatus, and the other in his own, the sensation conveyed to him will be as if a person were blowing into a tube at the opposite extremity.

Other trustworthy information is derived from inflating the drum. Cicatrices, thinned portions of the membrane, adhesions of it in various parts to the tympanic walls, insufficient mobility of the membrane from structural thickening, and various other morbid states, may be made apparent to us when we inspect it during a steady prolonged inflation. In making incisions or puncture of the drumhead, great assistance is rendered if it be distended outwards by this Valsalva method.

You may frequently see, in persons who have often resorted to the Valsalva experiment for improving their hearing, an altered position and curvature of the membrana tympani, and a perceptible thrusting outwards of the malleus-handle. The relaxed and sometimes thinned membrane lies as it were folded, and gives an undue prominence to the bone running down to its centre, looking somewhat like a piece of wetted bladder closing round an ivory knitting-needle. This is a condition truly termed by Wilde "collapse of the membrana tympani;" and now, if air be forcibly pressed into the drum, the membrane will become distended on one or both sides of the malleus-handle, like a pellicle, and the folded appearance is seen no more until the pressure is discontinued, when the membrane will sink back into its former state.

It sometimes happens that a patient becomes immediately and permanently relieved of a deafness of some duration, by thus forcing air into the drum. Such a case is related in my first lecture (page 14), and I might add a number of others in which great benefit or complete cure was brought about by this procedure alone. In young children who are very deaf, and who cannot be made to understand this method of overcoming the effects of their "infantile aural catarrh," it is certainly better in the onset to use the Politzer bag, instead of trying to teach them how to manage the Valsalva process. Some grown persons also are not able properly to inflate the tympanum, although their Eustachian tubes may be all the while permeable, of which fact the Politzer method or the catheter afford subsequently, evidence. As a means of diagnosis only, I consider the value of the Valsalva method to be on the whole as great as that of catheterism; but the latter we shall, I think, find to be more efficacious in the treatment, than applicable in the investigation of disease. Patients also, who readily obey instructions to inflate the drum by their own unassisted efforts, will not so willingly consent to have the catheter passed along their nostrils; and thus that instrument comes to be made use of chiefly in cases which obstinately resist the passage of air into the cavity of the tympanum by any simpler or less annoying methods. The "Valsalva method" of diagnosing the permeability of the Eustachian tubes, is however, by far the most frequently employed, and, in my opinion, will continue to be so for more than another century and a half. (Valsalva described it first in 1735.)

I have now to invite your attention to the very interesting and simple mode of treating Eustachian obstruction, called after the surgeon who, as I have before mentioned, was its discoverer, Dr. Politzer, of Vienna.

POLITZER'S METHOD consists essentially in blowing air into the cavity of the drum, by means of an elastic bag, while the Eustachian tubes are opened by their muscles during the act of swallowing. It was Mr. Toynbee who, in 1853, first demonstrated that the Eustachian tubes are in their state of repose, usually shut, and that they are opened by the palate muscles during every act of deglutition. Ten years later, Dr. Adam Politzer discovered that he could easily introduce air into the drum of the ear while the tubes were open at the instant the patient swallowed. This is effected by inserting a little way into the patient's nostril, a soft tube connected with an elastic pyriform bag.* The patient

---

* Messrs. Weiss and Son have modified Politzer's contrivance for me, by the addition of a valve, through which fresh air is drawn into the bag, thus obviating the necessity of withdrawing the latter from the nose after each compression.

first takes a mouthful of water: the surgeon then gently places the tube within either of the nostrils, and closes them both, by slight pressure with the finger and thumb of his left hand, sufficiently to prevent the return of air. Desiring the patient to swallow the water by degrees, the bag is quickly compressed with the surgeon's right hand, at each successive act of deglutition.

[Since these lectures were given I have still further improved Politzer's most invaluable appliance, by substituting a *nasal pad*, which is pressed *against* the opening into the nostrils, for the tube which he inserted *into* one of them. Mounted on a strong piece of covered copper-wire, are two air pads which can be so approximated or separated as to stop up conveniently, the nasal orifices. The metal serves as a handle for the surgeon or patient. Through each pad runs a hole, and these holes communicate with two short bits of Indian-rubber tubing joining into a single tube. Into this tube the pipe of the inflating bag is inserted, and the apparatus is thus complete. It can then be used, in the way above described, while the patient swallows.]

By this means repeated currents of air may be blown into the pharyngeal cavities, and this will force the warm air therein contained to enter and pass through both Eustachian tubes into the tympanum. The act of swallowing, as we know, opens the tubes, and the compressed stream of air is sufficient to overcome any obstructions of moderate extent in the tubes. Where the resistance to the full entry of air is not great, the patient feels suddenly a considerable pressure in the drum, and a noise is heard like a thunder-clap, or it is sometimes compared to "the explosion of a gun." I must caution you, however, against inferring too much from the sounds produced by this mode of inflating the drum.

Fig. 9.

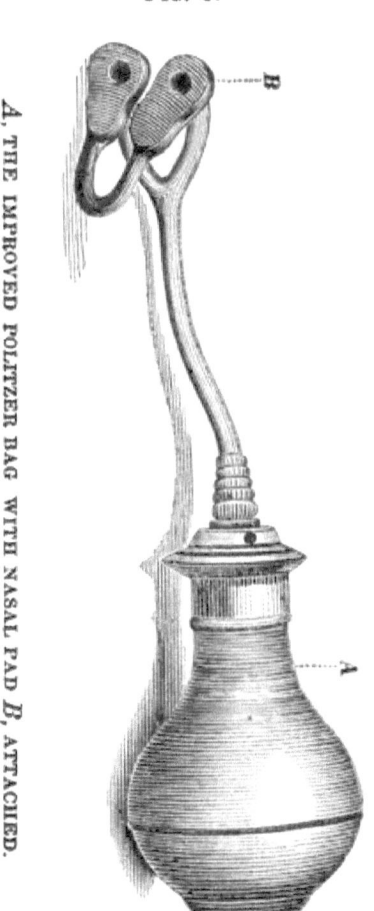

A, THE IMPROVED POLITZER BAG WITH NASAL PAD B, ATTACHED.

I see that statements have been put forth, which are not warranted by further experience in the use of this method. The otoscope placed in the meatus during this process rarely furnishes such satisfactory evidence with regard to the entrance of air, or the condition of the tubes and middle ear, as is afforded during the use of the catheter; for the noise made by the contraction of the muscles in swallowing the water, and that of squeezing the bag, mask the slighter sounds within the tympanum. The patient's own experiences are the best criterion of successful inflation. Of course, if a perforation of the membrana tympani exists, we become assured that such is the case, by the emission of a loud whistling or hissing noise, and the secretion from the middle ear is driven into the meatus, and sometimes issues from it. Moreover, if the membrana tympani (imperforate) be examined while the bag is being used, it may be seen to bulge outwards in different parts, generally at its posterior half.

You will perceive that the action of Politzer's method is very similar to that of the Valsalva process and to catheterism; but while more efficacious than the former, it is exempt from many of those objections which attend the use of the latter.

In treating young children, or nervous, irritable, debilitated persons, Politzer's method provides us a safe, easy, and painless means of removing those often-recurring obstructions in the Eustachian tubes which arise from aural catarrh; and hundreds of cases which formerly escaped efficient local treatment, we are now enabled to cure by the help of this excellent substitute for the catheter. Practitioners also, who are unaccustomed to the use of the latter instrument, will find in this new mode of inflation an invaluable aid to the treatment of aural affections depending upon catarrh, and even patients themselves are easily taught how to employ it.

FIG. 10.

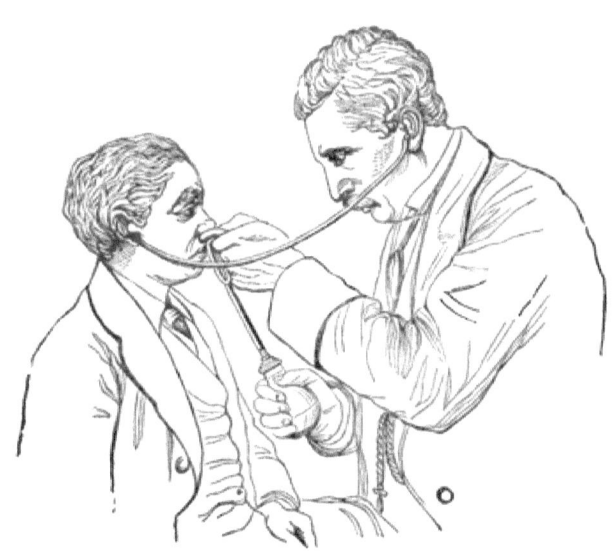

THE METHOD OF INFLATING THE TYMPANUM, WITH THE IMPROVED POLITZER BAG.

Considering the great frequency of this ailment in infants or young persons, and the facility with which the resulting deafness may be relieved before it has obtained such a hold upon them as to resist later remedial measures, it is not difficult to understand how deaf-mutism among our population might be diminished. We are continually informed that such or such a person, now deaf and dumb, only became so afflicted in infancy, or a little later, in childhood, after a severe attack of scarlet fever, measles, or some other infantile affection. If we can cure the deafness, the dumbness is prevented. A further value, I may here notice, attaches to Politzer's method, in the treatment of young children, inasmuch as it is not even necessary to cause them to swallow simultaneously with the compression of the bag. From the more ready distension of the infant's Eustachian tube, and the less resistance at its faucial orifice, the compressed air will generally pass from the nasal cavities into the middle ear without any swallowing motion being made.

Compared with the Valsalva method, Politzer's is the more efficacious, because it acts more powerfully on the obstructed and resisting passages without requiring the exercise of so much force on the part of the patient. We can, as I have stated, use it on children, and on adults who either cannot learn how to press air into the ear by blowing with the mouth and nose closed, or who (as in some instances) have superabundant growths of mucous membrane, folds, &c., which fall over the faucial extremities of the Eustachian tube. Nor do I find that the use of the inflating bag produces so much giddiness, as is sometimes felt when the patient brings into action the expiratory muscles of the chest in trying to exert the requisite pressure upon the orifices of the tubes in the Valsalva method. Compared with the catheter, the new method is of inferior value in the *treatment* of many of those intractable, long-standing cases where there are firm obstructions to the passage of air into the tympanum. Neither can it be a substitute for catheterism in the instances where a differing condition exists in the two ears, because air will rush into each simultaneously if the Eustachian tubes are pervious. Such a case now presents itself to my mind, in which, there being an insuperable objection on the part of a lady to the use of the catheter, I am obliged to forego any inflating operations to relieve a complete imperviousness of one Eustachian tube, because the other is too open, and the membrana tympani is perforated.* She hears tolerably well on this side, and fears any disturbance.

CATHETERISM OF THE EUSTACHIAN TUBE.

In England, the conviction appears to be yearly gaining ground that the Eustachian tube catheter is often a most indispensable means of diagnosis, and that many cases of deafness cannot be successfully treated without it. Toynbee's disbelief in its efficacy caused, somewhat generally, a less frequent resort to it, yet Pilcher, Yearsley and Wilde, among our own aural surgeons, as well as French and German aurists, had it in constant use. Toynbee's mistake in this respect has been already noticed. I have no desire to over-rate the value of Eustachian catheterism, nor

---

* I have since employed the catheter, and find that a *stricture* of the Eustachian tube exists. Until this is removed, the very high degree of deafness on this side, must remain.

to recommend its employment in cases where means less open to objection can be had recourse to; but I have no hesitation in saying that there is no other agent in the *treatment* of very numerous cases of catarrhal deafness upon which we may so confidently rely as the catheter. Its use, in fact, cannot be dispensed with. In *diagnosis*, I think, there is not such absolute need of it; and you seldom see me introduce the catheter, until convinced by the failure of other methods of inflation, that such a procedure is imperatively required.

I will first give you one or two facts connected with the history of the instrument, which will not be uninteresting, and then describe it, with its mode of introduction, its manifold uses in the treatment of deafness, and the effects produced upon the ear.

The existence of the Eustachian tube was known to Aristotle, but it was first described with tolerable exactness by Bartholomeus Eustachius in 1562, in his "Epistola de Auditus Organis." No attempt, however, to remove obstructions in the canal, was made until nearly 200 years afterwards, when, in 1724, Guyot, the Postmaster of Versailles, proposed to the Academy of Science at Paris, to inject it (the Eustachian tube), by means of a curved pipe introduced through the mouth. He is reported to have cured himself in this manner of a long existing deafness. In 1741, Cleland, an English army surgeon, apparently without knowing of Guyot's method, proposed the introduction of the catheter through the nose, the only practicable and certain way, and the only one now adopted. It may be interesting to hear his own words on the subject. He says: "If the person still remains deaf, the following instruments are made to open the Eustachian tube; if upon trial it shall be found to be obstructed, the passage is to be lubricated by throwing a little warm water into it, by a syringe joined to a flexible silver tube, which is passed through the nose into the oval opening of the duct, at the posterior opening of the nares towards the arch of the palate. By this catheter warm water may be injected, or it will admit to blow into the Eustachian tube, and so to force air into the barrel of the ear, and dilate the tube sufficiently for the discharge of the excrementitous matter that may be lodged there." No simpler or better description of the purposes of the instrument can now be given.

The catheter should be of silver, curved at the end, and bulb-pointed. A ring should always be affixed to the larger or funnel-shaped extremity, which will indicate the direction of the beak and the position of the point of the instrument when within the Eustachian tube. Catheters may be of various sizes, but three or four are sufficient for all cases. If they are made of pure silver, we may give them any curvature we wish. An elastic catheter is not so adaptable as a silver one. It may perhaps pass more easily through the nasal cavities, but there is more difficulty in hitting upon the mouth of the Eustachian tube with a flexible instrument acted upon by the hand at the opposite extremity. Experiments have shown that the stream of air upon the middle ear is not so strongly propelled through the vulcanized or elastic catheter, as through that made of an unyielding polished material. This may depend upon the fact that less pressure is made upon the faucial end of the Eustachian tube, or that a very flexible soft instrument does not enter it so far or firmly as the smaller silver one.

The steps of the operation may be briefly described as follows:—Place the patient in a chair opposite the light, with his head supported against the high back or by another person. I frequently place my own left hand round the back of the patient's head, for conveniently supporting it, as well as to keep it in position and prevent it from being turned to either side. Warm the catheter by dipping it in hot water, and introduce the curved bulbous end, with the beak directed a little downwards, into the inferior meatus of the nose; then quickly raise the whole instrument, and push it carefully but rapidly onwards along the floor of the nares, until it is arrested at the posterior wall of the pharynx. Then withdraw the catheter about half an inch or so towards yourself, at the same time turning its beak outwards and a very little upwards, looking at and feeling the ring at the end of the instrument between your finger and thumb, which indicates to you the direction of the beak, both being on the same plane.

Here I must pause, to criticise the above (my own) description of the operation, which I consider the must accurate, and also to remark that a dexterous and well practiced operator will frequently be able to turn the point of the catheter into the mouth of the Eustachian tube at once, without passing it so far back as pharyngeal walls and then having to withdraw it partially. This is that *tactus eruditus*, which nothing but sheer experience can give, but which is so difficult to describe. The parts over which the catheter passes vary in length and in calibre, and I therefore think it best for you to feel your way in the above mentioned manner. A mistake is commonly made at the stage when the instrument has reached the pharynx, by not withdrawing it far enough, and then in turning it the beak falls into the *fossa* situated behind the Eustachian tube.

LECTURE VI.

CATHETERISM—(*continued*).

GENTLEMEN,—A few precautions are necessary in manipulating with the catheter, and I will briefly enumerate them. Keep the catheter close to the floor of the nostrils until the pharynx is reached, for if care be not taken in this respect, the instrument may rise into the middle meatus of the nose, which is more sensitive and irritable than the lower passage. Moreover, you will probably fail to hit the orifice of the Eustachian tube. Observe,—if the catheter has entered the middle instead of keeping in the inferior meatus, it will not form a right angle with the plane of the face, but an oblique one. Use no force whatever, especially in rotating the point of the instrument near the pharyngeal orifice. It must be recollected that in the greater number of cases where we employ catheterism, the naso-pharyngeal mucous membrane has been more or less diseased, and, consequently, any want of delicacy in handling the instrument may easily produce a rupture of its walls. Should the point of the catheter perforate the mucous membrane, or pass through any thinly-cicatrised ulcer, as soon as you inject air you may produce emphysema of the subjacent cellular membrane, and blow up your patient like a joint of veal. My teacher, Mr. Pilcher, the most eminent aural surgeon of his day, in passing the Eustachian catheter upon a patient more than twenty-five years ago, hit unfortunately upon the old cicatrix of an ulcer caused by small-pox. When air was pressed into the catheter, the cellular membrane of the whole side became emphysematous, and for a time rather dangerous symptoms ensued. This woman afterwards became a most excellent servant in Mr. Pilcher's house, and during several years of my residence with him, she regularly presented herself twice a week for me, then a pupil, to catheterize her Eustachian tubes, so much benefit did she derive from it. This, after such an accident, I cannot help remarking as a very uncommon instance of confidence and good sense.

Hæmorrhage sometimes occurs from the irritable and congested mucous membrane; but cold water or a simple astringent application of alum will easily control it. I find that though patients may be excessively nervous and timid on the first occasion of passing the catheter, they will, when it is repeated on subsequent occasions, be calm and quiet. The operation is comparatively speaking, painless, and you yourselves often see patients returning again and again to submit to it, finding real and increasing benefit to their deafness. The same occurs in private practice, and excellent indeed must be its effect, where hearing, defective for many years, is thus not only prevented from becoming worse, but even improved. It has fallen to my lot to pass the Eustachian tube catheter, certainly many thousand times, without as yet any of the accidents above noticed having occurred.

When the catheter is ascertained to have been properly inserted into the mouth of the Eustachian tube, air may be blown in through it, by means of a gutta-percha bag, a syringe, or through a piece of tubing placed in the surgeon's mouth. This last method of introducing your own breath into the nasal cavities and throat of your patient seems to me an objectionable proceeding, and I merely mention it because such a mode is the only one suggested by the late Mr. Toynbee, and is figured in his work, page 203. The syringe, provided with a nozzle accurately fitting the larger orifice of the catheter, will propel either warmed air, medicated vapours, or fluids, with as powerful or gentle a force as may be desired. Let me here caution you against the use of air-presses and pumps, or of any compression-apparatus which you cannot regulate with your own hands. I never employ any mechanical means for forcing in air other than the syringe or elastic bag. The former, by being first filled with hot water, which is subsequently discharged, will convey from its heated barrel air quite sufficiently warmed into the cavity of the tympanum; while the latter (the bag) I chiefly use for injecting medicated vapours through the catheter. A small bottle containing the agent, the vapour of which is designed to be introduced into the drum through the Eustachian tube, is either held in the hand or placed in hot water; the usual stopper of the bottle is substituted by a cork or another glass stopper with two holes in it, one hole being for the tube connected with the compression-bag, and the other for the entry of air. These I find sufficient for all purposes, and the numerous additional appliances, such as the frontlet-band, spectacle-forceps, nose-pincers, and the like, for holding the catheter tightly in the nostrils, have never met with approval at my hands. You may see them illustrated in most works on aural surgery; but I am satisfied you will never wish to employ them, if you witness for yourselves how easily and conveniently they may be dispensed with.

The practice of injecting fluids, medicated or otherwise, for the purpose of washing out the cavity of the tympanum, is so important for good or evil that I shall advert to it in an especial manner when considering the local treatment of the various parts of the middle ear affected in catarrh. It is sufficient now to say that fluids must only be pressed into the tympanic cavity with the greatest precaution, and that the main object in using them at all is that of altering the condition of the lining membrane when diseased, and restoring it to a healthy state. In employing such topical applications to the mucous cavities, not only of the drum, but necessarily and simultaneously to the large expanse of the mastoid cells, never forget that these parts cannot be completely drained of the injected fluid, unless there exist an aperture quite at the lower margin of the membrana tympani, because, as I have before pointed out, the Eustachian tube does not make its exit from the floor of the tympanum, but nearly half-way up.

Lastly, it may be here mentioned, that the catheter provides a means for introducing bougies into the Eustachian tube, when there is reason to believe that we have to deal with a stricture so firm as to resist all other modes of dilation. In such a case where restoration of its patency cannot be effected without using severe mechanical means, the catgut bougies, and *laminaria digitata*, or sea-tangle tents, are employed; but even greater caution is required in the introduction of such extremely irritating foreign substances into so delicately organised a part as the middle ear. This method of treatment has been of late far too frequently resorted to.

The above are the *uses* of the catheter. It is, in short, an instrument which opens up to us the Eustachian canal and the cavity of the middle ear; and enables us to restore air, and apply remedial agents to parts which are otherwise inaccessible except there be an aperture in the membrana tympani, which will of course expose the drum externally.

The *effects* of catheterism now only remain to be noticed.

If a stream of air be forced through the instrument into the cavity of the tympanum, and one end of the diagnostic tube, or otoscope, be placed in the meatus of the patient, the other in our own, we hear a kind of *thud*, or cracking sound, caused by the bending outwards of the dry elastic membrana tympani; and this is followed by a continuous rustling noise, termed by the French, "bruit de pluie" or falling of rain, and by the Germans, "auschlage geraüsch," knocking sound. These sounds are produced if the Eustachian tube and tympanum are of normal size, and the lining membranes of each have a healthy degree of moisture; but if there is obstruction of the former, or a diseased condition of the latter, we are furnished with certain valuable signs, which (taken together with the history of the case and the appearances of the membrana tympani), will indicate either with tolerable accuracy. For example, if the sound be harsh, full, or sharp, it may be inferred that the mucous membrane is particularly dry, and the Eustachian tube freely permeable. Such a condition is of frequent occurrence with old people, and in certain chronic inflammations. If the sounds be squeaking, interrupted, or piping, they indicate some obstruction or diminished calibre of the Eustachian tube, and a morbid state (generally from the thickening) of its lining mucous membrane. An ear practised in auscultation will also be able to determine whether the sounds be near to or far from the examiner's ear, and thus distinguish in diagnosis whether there be partial stricture of the Eustachian tube, or whether the mucous membrane of the tympanum itself be constricted by induration or adhesion of its walls. A gurgling moist sound will intimate to the surgeon's ear that there is an excess of mucus or other fluid secretion in the drum; or, finally, we may be assured (as by the Politzer or Valsalva methods) that there exists a perforation in the membrana tympani, if the air be whistled or hissed out plainly into the otoscope, or if a few drops of pus or mucus be forced into the meatus. This " perforation murmur " can never be mistaken. The meanings of all these various sounds as made out by auscultation, will be matter for more detailed study as we go on; and as they are now arranged in some degree of order, we shall be much assisted in diagnosis by becoming thoroughly familiar with them.

If the catheter can thus help us to recognise certain morbid conditions of the middle ear and Eustachian tube, it will obviously afford some indications for treatment; but as I have before remarked, its value is chiefly therapeutic, especially in catarrhal affections. In *diagnosis*, we may often substitute the Politzer or Valsalva method for the catheter, and only have recourse to the latter in doubtful cases, and when its employment is desirable at the same time as a therapeutic agent, for introducing air, vapour, pulverised fluids, bougies, &c. In the *treatment* of diseases of the middle ear, its efficacy is most decided.

In the above enumeration of the effects produced by inflation of the drum cavity, we have mainly confined our attention to the *sounds heard* when the ear is auscultated. After using the ear douche, however, we must never omit also to inspect the membrana tympani, in order to convince ourselves by the sense of sight that a certain result has occurred upon this part of the organ, because if we examine the drum-head in certain morbid conditions, while air is being blown in through the catheter, or immediately afterwards, we shall see that it has been moved outwards towards the auditory canal, either wholly or in part, according to the amount of the obstruction or nature of the adhesions met with.

If there was an accumulation of mucus in the cavity of the tympanum, as previously conjectured, both from the appearance of the external membrane and the moist gurgling sound in catheterism, the air douche will have dispersed it about in various directions,—some will have been driven into the mastoid cells, and some will have passed into the throat through the Eustachian tube. The more solid agglutinations and adhesions of inspissated mucus clinging to the membrane, which have perhaps remained for a long time within the tympanum, and produced more or less severe deafness and troublesome tinnitus, will by the air douche have been loosened, and the membrana tympani permitted to resume a more normal position, so that a marked change in its form, curvature, and colour may be at once perceptible. As such conditions as these are among the most frequent results of aural catarrh, and can be generally relieved, and sometimes cured, by the introduction of the catheter and a few air douches, we are quite justified in promising to our patients not only temporary improvement in their hearing, but that probably a lasting cure may be effected by repeated catheterism of the Eustachian tube. It is, as you have had pointed out to you among the hospital aural patients, quite *possible* to break up adhesive processes going on or already established among the various delicate osseous and membranous contents of the tympanum, by warm air or warm water injections; and the severe deafness which was entirely owing to the rigidity of the ossicular chain and perhaps impaction of the stapes in the fenestral opening, may thus be overcome through the use of the catheter.

An interesting case is just now under my observation and care, in which, during the inflation of the ear, portions of the membrane become quite protruded in the form of two vesicles on each side of the malleus, and while they remain in this state, the hearing improves in a marked degree; but as soon as the air-pressure from within ceases, the bulgings fall back. In this case I suspect, indeed I think I have once or twice become convinced, that a small perforation exists in the mucous and fibrous layers of the membrana tympani, and occasionally in the external or cutaneous layer; but that the last mentioned, the thinnest, cicatrizes and permits the air to pass under it. We might, not improperly, term this condition *emphysema of the membrana tympani*. In the same patient there are also decided evidences of adhesion of the malleus to the promontory. Operations have, however, increased the hearing power from two inches to two feet, and relieved the constant tinnitus.

Lastly, with regard to the effects of catheterism, it will be apparent that the pressure of air within the cavity of the tympanum, when thus introduced through the Eustachian tube, may, in some abnormal conditions of the drum-head, be exerted unequally upon the fenestral openings. This consideration has a most important bearing upon the prognosis of a case, for if we find the adhesions very firm, and the membrana tympani almost or quite persistently immovable under the air douche, we shall not be warranted in holding out expectations of either temporary or permanent cure of the accompanying deafness; and our prognosis must therefore be unfavourable.

Let me recall to you what has previously been said on the anatomical relations of the two fenestral openings to the membrana tympani. The chain of bones is fastened at one end to that membrane by means of the handle of the malleus, and at the other end (at the fenestra ovalis) presses against the labyrinth-fluid by means of the base of the stapes. Every pressure inwards of the external membrane must therefore simultaneously thrust inwards the stapes, which will so influence this fluid (perilymph) as to cause a corresponding thrusting outwards of the membrane at the fenestra rotunda. This compensation movement takes place, you are aware, because the perilymph is confined within unyielding bony canals, and could not be affected by vibrations of air, were it incompressible. If from any cause the stapes cannot be moved freely in and out of its oval opening, no corresponding stretching and relaxation of the membrane at the round opening can occur, and if in such a case air be forcibly blown into the drum through the Eustachian tube without pushing outwards the membrana tympani, the whole ossicular chain will remain immovable, and the labyrinth fluid at the oval opening will be uninfluenced by any withdrawal of the stapes. Very different, however, is the result upon the delicate membrane at the other end of the labyrinth, the round cochlear opening. The pressure of the condensed air is here very great, and may sometimes cause most unpleasant giddiness. Now it is plain that adhesions of the membrana tympani to the walls of the tympanic cavity, or anchylosis of the stapes to the fenestra ovalis will most materially restrict the excursive power of the drum-head, if not altogether hinder it, and consequently no relaxation of the important secondary membrane (membrana fenestræ rotundæ) can be effected; it remains inelastic and unvibratile, and no improvement to the hearing can be anticipated while it continues so. Should, however, these adhesions, and the rigidity of the ossicles, be broken up by the repeated and carefully regulated introduction of streams of air, the lost elasticity of the membrane of the fenestra rotunda will be restored, and the deafness, so far as it depended thereupon, cured.

LECTURE VII.

AURAL CATARRH.

GENTLEMEN,—We now commence the study of Aural Catarrh in all its varieties and consequences. This being the most frequent of all affections of the ear, and the commonest cause of deafness, whether in childhood, middle life, or old age, the subject will deserve our special attention. In order that the description of the disease and of its treatment might not be interrupted when once entered on, I gave you in the foregoing lectures a brief account of all which relates to the mode of investigating by mechanical appliances, and diagnosing the ailments of the middle ear and Eustachian tube; and we have likewise (necessarily in a very cursory manner) taken a view of the anatomical relations of the parts most concerned in aural catarrh. The membrana tympani, from its being so accessible to inspection, and also because it is so generally implicated in most ear diseases, engaged a correspondingly larger amount of anatomical notice; and its appearances in health were dwelt upon at some length, to the end that the minute changes effected by catarrhal disease might be more precisely recognised.

Authors have written upon diseases of the tympanum under various appellations, such as "Catarrhal Otitis," "Chronic Internal Catarrh," "Mucous Accumulation in the Tympanum," "Inflammation of the Eustachian Tube" (without mention of the tympanic cavity at all), "Engorgement and Obstruction of the Cavity of the Tympanum," "Milder Form of Acute Otitis Interna," &c., &c.; and Dr. Kramer has described it under the heading of "Inflammation of the Mucous Membrane of the Middle Ear, with Accumulation of Mucus." When the membrana tympani alone appears to be affected, aural catarrh has been designated by Sir William Wilde, "Subacute and Strumous Myringitis;" and again, when the Eustachian tube presents the most prominent symptoms of catarrhal affection, the reader is referred to a chapter or two on the diseases appertaining to this particular part of the middle ear.

Such divisions of the subject appear to me very unsatisfactory; for it is perfectly useless to attempt to separate, in description, the catarrhal diseases of the Eustachian tube and mastoid cells from those of the tympanic cavity itself; because all the affections now under notice have their origin in the mucous layer of some of the parts of the middle ear just mentioned, and will assuredly sooner or later extend over the whole mucous layer of the cavitas tympani, thus becoming common to all.

They may, therefore, be simply and correctly designated as "aural catarrh."

It is evidently impossible to limit inflammatory action, whether acute, subacute, or chronic, to the particular structure in which it originally commenced; therefore neither an anatomical nor a strictly pathological nomenclature of catarrhal disease is quite feasible. The peculiar symptoms and effects of catarrhal inflammation upon the various parts composing the whole of the middle division of the ear, will in the following synoptical arrangement, be consecutively described and studied.

> First. SIMPLE AURAL CATARRH, or Catarrhal Inflammation of *the mucous membrane of the cavitas tympani, membrana tympani, Eustachian tube, and mastoid cells.*
>
> This form of catarrh may be divided into *acute and chronic.*
>
> Second. PURULENT AURAL CATARRH, OR OTITIS, also *acute and chronic.*
>
> Third. OTORRHŒA, AURAL POLYPI, &c., or the results of purulent aural catarrh.

Acute Catarrhal Inflammation of the TYMPANUM may consist in mere hyperæmic swelling of the mucous membrane, with slightly increased secretion of mucus; or it may extend into a most dangerous and even fatal disease.

The first form is sometimes very transient, passing off in a few days, without leaving any impairment of function behind. The *causes* of such a mild and temporary deafness are usually supposed to be "catching cold," getting the body or feet wet, influenza, or the like. The *symptoms* are described by the patient as "having a cold in the head," stuffing of the nostrils, heaviness over the brow, sore-throat, singing or tinnitus in the ears. The slight impairment of hearing which accompanies these symptoms is occasionally relieved by blowing the nose, coughing, sneezing, &c. The parts involved belonging to the ear are, the Eustachian tube, especially in its faucial portion, and the whole cavitas tympani; and we also generally observe that such conditions of these parts as are above described are associated with congestions or catarrhal inflammation (as they are termed) of the nasal passages, fauces, bronchial tubes, and sometimes of the lungs.

This mild, though acute and sudden, catarrhal affection (which might more properly be named *congestion*) produces seldom any decided alteration in the appearance of the membrana tympani or of the auditory canal. Occasionally there may be a somewhat reddish tinge mingled with the grey colour of the membrane, resulting from the injection of the mucous layer within; and sometimes the transparency of the membrana tympani permits the congested interior layers of mucous tissue, with a large amount of accumulated secretion, to be distinctly seen.

In this latter form, which may be considered as the *second* degree, or a state between simple congestion and progressive inflammation, hearing is impaired, singing in the ears becomes annoying, a sensation of fulness is experienced, and when the drum is inflated by the Politzer method or catheter, the moist gurgling sound indicative of the presence of fluid within the cavity of the tympanum, is plainly heard through the otoscope or by the surgeon's practised ear without the assistance of that instrument.

In the *third* degree, when the inflammation is higher, the symptoms are more general. The throat participates in the attack; its mucous membrane is tumid, thickened, and severely injected, and swallowing is often painful and difficult. The acts of coughing, sneezing, or blowing the nose increase the pain by the sudden and forcible expulsion of air through the constricted and inflamed Eustachian tube. Cracking sounds and a feeling of bursting are complained of in the ears, or the drum cannot be inflated at all, owing to the complete closure of the tube. The patient's own voice is described as "reverberating," or "hollow." The tuning-fork placed on the vertex of the head, will, in consequence of its vibrations conveyed through the cranial bones, being prevented from easily escaping out of the auditory meatus, be more distinctly and longer heard by the patient than by the surgeon.* I have frequently tested the power of hearing sounds thus transmitted in those suffering from catarrhal inflammation of the middle ear; and when, as compared with my

---

* See page 37 for explanation of these phenomena.

own, the perception of them is enhanced, or the same as mine, there is tolerable certainty that the important fenestral openings in the labyrinth-walls have as yet escaped any considerable implication. It may also be inferred that the portion of the tympanic mucous membrane which lines the internal walls of the cavity is neither rigid nor much thickened; for when it is severely affected by these catarrhal processes, I have almost invariably found that fixity of the base of the stapes in the fenestra ovalis, as well as more or less impairment of vibratility in the membrane of the fenestra rotunda cochleæ, is the result. The tumefied and thickened mucous membrane binds down these elastic oscillating structures, and disables them from influencing by their rapid wave-like movements the fluid contents of the vestibule and cochlea, in which the filaments of the auditory nerve are, as you recollect, suspended. So that, when there is little or no perception of sound through the cranial bones, and from the history of the case and other general symptoms there is no reason to suppose that the auditory nerve itself is diseased, the deafness in catarrh may be attributed to excessive pressure upon the fenestræ, and consequently upon the contents of the labyrinth, by hyperæmic, unyielding mucous membrane. Rigidity of the whole ossicular chain, and exudation of lymph in the tympanic cavity, will likewise obviously cause the mobility of the stapes in its fenestra, and the vibratile power of the membrane of the fenestra rotunda, to become impaired; and in addition to these serious hindrances to hearing, the membrana tympani may have become so swollen and thickened by organised deposits, hæmorrhagic exudations, &c., upon its inner mucous layer, that the cavity of the tympanum is reduced in size. This condition superadded, will assist in wholly stopping the vibration of the auditory ossicles, but would not of itself, I think, prevent the tuning-fork from being heard through the bones of the head, if the mucous membrane at the inner wall of the drum abutting upon the labyrinth were not implicated in the disease.

I have been thus particular in drawing attention to the use of the tuning-fork, as a modern and an additional means of distinguishing catarrhal disease, occurring in the cavity of the tympanum, and causing, by intra-auricular pressure, impairment of function, from those ailments in which the nervous apparatus is primarily affected, because you might with much reason inquire during the examination of a patient by whom the tuning-fork is not heard when placed on his forehead, " Why may we not consider the case as one of *nervous* affection, originally and chiefly ?" In my third Lecture it was remarked that there is a limit to this mode of investigating, and now you have evidence that the objective symptom must never be viewed alone, but in connexion with others, and also especially with the history, cause, and duration of the disease. By practice, and attentively observing and analysing the cases brought before you, I am sure you will not find it difficult to arrive at correct conclusions on this most important ·point. I need not tell you how essentially necessary it is that you should be able to inform your patient whether the severe deafness depends upon a morbid condition of the drum cavity (which in simple acute catarrh is generally transient though urgent), or that you are compelled to infer from this objective mode of examination the infirmity to be "nervous," in which case recovery is not to be expected.

I may here digress for a moment, and observe that it is somewhat consolatory to deaf persons, and may induce them more readily and earlier to apply for professional assistance, if they learn that with the increase of our knowledge of such morbid processes as take place in the middle ear (that is, externally to where the nerve is expanded), and the improvement in examining and distinguishing its various affections, the number of (so called) "nervous" cases has wonderfully diminished. Those which were formerly regarded as incurable, because their pathological conditions were not recognised, and therefore could not be removed by treatment, are now, by the progress of scientific aural surgery, brought under some degree of control; and both in objective modes of investigating nervous complaints, and in treating the diseases of other parts of the ear, mistakes are less frequent than formerly, and cures more numerous.

Acute catarrhal inflammation of the MEMBRANA TYMPANI.

The membrana tympani is seldom affected alone, but generally by extension from the mucous membrane lining the cavity. The acute form of this ailment is more uncommon than the chronic, but requires energetic treatment to prevent it from lapsing into the latter. Many persons can trace their difficulty of hearing to acute attacks of catarrhal inflammation in early life, when they were considered to have merely an "ear-ache" from cold, "not of much consequence," or they may have suffered repeated catarrhs of the ear, and recovered their hearing without any special local treatment. We now, however, know that when patients have once been the subjects of aural catarrh, they are predisposed to the return of deafness, which in course of time becomes not only severe, but permanent. The cause of this recurrence of deafness is easily made clear, both from anatomical and pathological considerations. I have already stated that in *tympanic* catarrh a swelling and thickening of the mucous membrane occurs, causing adhesions and connexions between parts which should be free to vibrate; and it will be observed that such attachments most frequently occur between the membrana tympani and promontory, between the same and the ossicles, and between the ossicles (the stapes very generally) and fenestra ovalis. When the cavity of the drum is thus diminished in size by hyperæmia, by thickened lining membrane, adhesions, and abnormal connexions, it is very evident that any slight additional swelling or congestion, such as results from a "cold in the head," or the sudden application of cold externally to the membrana tympani, by draughts of winter air, &c., will almost obliterate the already narrowed space of the drum-cavity, and most sensibly impair the hearing. In an analogous condition of the eye, where the iris becomes adherent to the capsule of the lens (constituting senechia posterior), proper vision is hindered in consequence of irregularity in the muscular action which is necessary in accommodating the eye for sight at varying distances. In the same way, the two muscles of the ossicles (the tensor tympani and the stapedius), which undoubtedly regulate their movements, and accommodate the ear to receive sounds of all characters and degrees of intensity, will when confined and limited in their action by adhesive bands, the result of catarrhal inflammation, render the motions of the membranes of the tympanum and the labyrinth irregular, uncertain, and inharmonious. It follows that even if the deafness be not severe, the power of hearing ordinary conversation or several sounds at one time is much diminished, and many patients thus affected can only hear when they listen attentively to the person talking. This symptom is not unimportant, since it indicates very plainly the seat of the mischief. The physiological explanation of the inability to distinguish the various intonations of the human voice when the action of the tympanic muscles, especially the stapedius, is impeded by the thickened mucous membrane of the drum-cavity, was given in my fourth Lecture (page 57). Whenever, then, you see a patient dull of hearing, vacant in look, and only when he listens attentively, capable of entering into conversation and distinguishing what is said (that is, only when he exercises the stapedius muscle by an effort of the will), it may confidently be assumed that the accommodating power of the ear is defective or lost in consequence of congestion and thickening of the lining membrane, plastic exudations or adhesions in the cavity of the tympanum.

The description of these symptoms and states of the middle ear, and of their liability to recurrence, will impress on your minds the necessity of preventing by timely treatment, their becoming the source of permanent deafness. If patients are unwilling to submit to remedial measures, expostulate with them, and explain the probable consequences of their refusal. They will not " grow *out* of" their ailment, as medical men are almost daily reported to me to have said, but *into* it.

The *pain*, varying according to the degree of inflammation, and the layers of the membrana tympani affected, may amount merely to uneasiness, intermitted or continuous, or to intense agony. Acute attacks of catarrhal inflammation of the drum and its membrane are more common with children than in the adult. They are apt to occur after exposure to cold, blasts of driving wind and bathing, generally coming on suddenly in the night, and increased by the recumbent position, and pressure of the ear on the pillow, which seem to favour congestion of the membrane. In the adult, the symptoms are more formidable in character; and persons of a rheumatic or gouty diathesis appear to be more prone than others to catarrhal inflammation of the membrana tympani, owing, I believe, to its peculiar texture, in which fibrous membrane forms so large a part. It is not always possible to define the symptoms of an acute catarrh originating in the cavity of the drum and invading by extension the fibrous laminæ of its membrane (well termed by Wilde *myringitis*), but the pain which accompanies them is usually more severe and lancinating than when merely the internal mucous or external cutaneous layers are implicated. It is often compared to a knife running into the brain, is increased by swallowing, coughing, sneezing, blowing the nose, &c., and becomes occasionally so excruciating as to induce delirium. The pulsation of the carotid artery (which lies in such close proximity to the inflamed structures) is not only distinctly heard in the ear, but each throb is felt, adding to the general excitement and suffering. The whole of that side of the head on which the inflammation fixes is acutely painful to touch or motion (even the eye, teeth, temple, &c.); in fact, the severity of the pain experienced is a test of the amount and extent of the inflammatory action; so that if the majority of the above-mentioned symptoms be present, we may conclude the whole cavity of the tympanum, as well as the several layers of the membrane, to be involved. Sometimes fatal consequences ensue from inflammation extending to the brain, but such cases, together with those terminating in suppuration (constituting *purulent aural catarrh*), will be more particularly referred to when that subject comes under notice.

Deafness comes on contemporaneously with the pain, or soon afterwards, but if only one side be attacked the patient will not invariably be sensible of the diminution or loss of his hearing. In rare cases at the commencement, an exaltation of the auditory sense occurs (hyperacusis, analogous to photophobia in ophthalmic disease). *Tinnitus aurium* of all kinds is usually coincident with the painful seizure, and the greater the nervous excitement, and higher the degree of inflammation, the worse will be these noises.

To these local symptoms must be added certain well marked constitutional ones, such as impairment of the functions of the brain, distress, anxiety and depression of mind, giving rise to the worst forebodings, restlessness, sleeplessness, or in the severest cases, delirium, and symptoms of cerebral disease. An example of this will be cited by-and-bye.

Appearances of the membrana tympani and meatus externus in acute catarrh.

On account of congestion of its mucous layer, and opacity of the fibrous and dermoid layers, the *colour* of the membrana tympani undergoes several modifications. At the commencement of the attack, when the symptoms of acute catarrh or of myringitis are somewhat severe, the external surface has a peculiar glistening appearance, which gives the uniformly reddened membrane a polish like a "copper plate." This state does not last long, for the epidermic layer soon becomes infiltrated, and the shining appearance gives place to a dull pinkish hue which may vary to all the shades of red, according to the degree of vascularity, which of course corresponds to the greater or less amount of inflammation, and the component parts of the membrane engaged. Sometimes, as in the coats of the eye, new blood-vessels seem to start into existence, and encircle the membrane with a zone of vivid red, as seen in iritis, or on the borders of the cornea in corneitis.

The arteries running down the malleus-handle are especially marked out at the onset, but afterwards their separate ramifications cannot be distinguished, owing to the increase of vascularity, and the malleus-handle itself will be lost to view in the generally swollen and reddened mass. The auditory canal is not much affected in the earlier stages of the disorder, but later it becomes injected, dry, glistening, and swelled, especially at the part contiguous to the drum-head, so that the usually well-defined boundary between the two is not recognizable. *Ecchymoses*, or irregular hæmorrhagic spots are seen on the surface of the membrane, or within the drum if the membrane remain sufficiently transparent. *Vesicles* are also observed, caused by the separation of certain portions of the epidermoid layer, through effusion of serum, which look like pustules, and although these subside with the acute inflammatory action, we can never accurately decide whether the case is to develop into merely a simple catarrh, or to become aggravated into the purulent form.

LECTURE VIII.

SIMPLE AURAL CATARRH—(*continued*).

Gentlemen,—*Acute Inflammation of the Eustachian Tube.**—It has been proved that the mucous membrane lining the tube is an extension from that of the pharynx, and that it passes into the tympanum and mastoid cells, without any solution of continuity or line of demarcation whatever.

There needs, consequently, no further demonstration to convince you that when a severe affection of the fauces, soft palate, uvula, tonsils, or any portion of the naso-pharyngeal mucous membrane exists, the Eustachian tube must almost inevitably participate in the congestive and hyperæmic condition. Moreover, when we consider the anatomical relations of the tube and mastoid cells to the cavity of the tympanum, we cannot be surprised to find that in these parts, at the further end as it were, of

---

* For an outline of its anatomy see page 63.

the same system of mucous surfaces, a similar condition will generally co-exist, and the subjective symptoms will be much like in character to those enumerated in the last lecture as belonging to catarrhal inflammation of the tympanic structure itself. In the following description it will be impossible always to avoid repeating details and symptoms common to inflammation of the part last named, because the diseased action, though it may have begun elsewhere, sooner or later extends to the drum and its contents. In this purely mechanical way, therefore, we have a thorough and complete aural catarrh established from simply "catching cold," "getting the feet wet," "having a sore throat," &c., according to the patient's description.

The appearances of the different mucous surfaces in inflammation, will vary according to the particular portion affected — for there are numerous morbid changes taking place—but if rhinoscopy were more practicable than it is while our patient is the subject of acute pharyngeal inflammation, we should be able to describe the pathognomonic evidences with greater precision. If the throat be looked into (and it should always be done) when either the Eustachian tube, the cavity of the tympanum, or its external membrane is in an inflammatory condition, we notice its mucous membrane to be red, tumid, velvety, spongy-looking, and infiltrated; or with even in the mildest cases, severe injection of the blood-vessels. The nose and frontal sinuses partake in the unhealthy condition of the adjacent tissues, and a feeling of "stuffing" in this part is present, together with the characteristic open mouth and faucial, instead of nasal, breathing so commonly observed. In this form the ailment attacks chiefly children and young or middle-aged persons; and the light-haired, fair-complexioned, debilitated, or scrofulous are more liable to it than others. There is a want of power over the muscles of deglutition; swallowing is therefore difficult, and accomplished with pain, which is aggravated by every motion of the throat; for it will be remembered that the muscles concerned in deglutition are also muscles of the Eustachian tube, and that their fibres interlace themselves among the glands and neighbouring mucous surfaces of the soft palate. By reason of their intensely inflamed and painful state, these muscles of the palate and Eustachian tube are rendered incompetent to perform their proper function of drawing apart the walls of the tube each time the patient swallows. The lining membrane of the Eustachian tube being already much congested and thickened, and perhaps secreting mucous too abundantly, the patient experiences a peculiar sensation of fulness and occasional "bursting" in the drum of the ear, which most of us have been conscious of when suffering from influenza or catarrh.

In such patients the voice sounds hollow, in proportion to the amount of obstruction in their Eustachian tube and tympanum; and this is a symptom peculiarly marked if both ears are affected, or if the unaffected one be closed with the finger. On the abatement of the more severe symptoms (such as fever and pain) already noticed as common to tympanic and tubal inflammation, the dull, heavy, and stuffed feeling in the ear, head, and throat, with deafness, will perhaps remain. Sometimes, however, relief to these sensations is afforded by dislodgment of the mucous secretion from the guttural orifice of the tube, which is accompanied by a loud "crack," or report, in the ear. This "pop" may occur suddenly during sneezing, yawning, blowing the nose, vomiting, or some sudden respiratory effort, and is generally the prelude to improvement or complete restoration of hearing. The amendment, however, may be only temporary, and the hearing may be lost again. This makes it necessary that I should explain to you more in detail how the results of acute catarrh of the Eustachian tube (more especially in reference to its faucial end), may not only damage the patient's hearing at the time, but in spite of *general* treatment occasion future severe or perhaps incurable deafness.

The lower portion of the tube being implicated in almost every "bad cold," its closure follows, either in consequence of the thickening of its walls, or from hyper-secretion of mucus within the very small channel which it naturally is. The walls of the tube thus become more or less firmly agglutinated, and *greater muscular action is required to draw them apart.* Now, when these muscles which move the palate in swallowing, and by their contraction pull upon the walls of the tube and open the passage into the (otherwise closed) tympanum, are hindered in their action, air can neither enter nor recede from that cavity. Thus, one of the primary requisites for perfect audition (viz., the equalization of density between the atmospheric air, and that contained within the drum), is absent. No interchange occurring, the air which was in the cavity of the tympanum gradually disappears in the course of a day or two (probably by absorption), or at all events it becomes rarefied or exhausted, being doubtless influenced thereto by the increased heat, and congestive swelling of the mucous membrane lining this occluded chamber.* The immediate effect of this unnatural condition is to produce increased concavity (as seen from the outside) of the membrana tympani, a forcing inwards of the chain of ossicles, and undue pressure upon the contents of the labyrinth, causing the hearing powers to be seriously diminished. In addition to this mechanical effect upon the important structures of the internal ear, by the impaired function in the Eustachian tube-muscles, and the consequent impermeability of the tubular passage, you will not be surprised to learn that other abnormal symptoms are soon manifested in the cavity itself, such as have been previously detailed. In severe cases, exudations from the acutely inflamed mucous membrane rapidly follow; viscid, tenacious mucus is abundantly poured out; and this being prevented from escaping through the Eustachian tube (its natural

---

* For physiological explanation of the phenomena caused by rarefaction and condensation see also page 66.

outlet), settles upon the delicate little bones and their muscles, and impedes their vibratility and adjusting action, throwing their mechanism out of order, or at the very least rendering their motion irregular and out of harmony;—because of the disproportion thus developed between the power possessed by and the work demanded of these little "accommodating muscles" of the ossicular chain. The rarefaction or exhaustion of tympanic air, together with the congestion (however slight) of the lining membrane, which would produce no appreciably injurious effect upon a cavity of normal size, will in one that has been narrowed as just described, lessen the acuteness of hearing in a most sensible degree.

The further results of continued closure of the Eustachian tube, as well as the various forms of adhesions and attachments which are developed from the contact of mucous surfaces within the middle ear, belong more properly to the subject of *chronic* catarrh of the tympanum, and will be hereafter fully noticed, as indeed their importance demands, likewise other morbid conditions causing tubal obstruction, such as *thickening* or *relaxation* of the mucous lining membrane.

Occlusion of the Eustachian tube, with some of the before described consequences, may be observed in some patients to accompany every cold in the head, influenza, or bronchial attack, and if the hyperæmia and tumefaction only last a short time, the middle ear will generally recover its functions as soon as the communication between the throat and drum-cavity is restored. Equilibrium of air being established on each side of the drum-head, this thin yielding membrane will resume its normal position, form, and elasticity. Freed from undue pressure inwards of the membrana tympani, and through it, from pressure of the last bone in the ossicular chain against the tender structures of the labyrinth, the patient will lose the sensation of fulness in the ears, and the distressing noises. In some of the severer cases, particularly in children, it is difficult at the commencement to distinguish an acute attack of aural catarrh, with completely obstructed tube (which of course aggravates the subjective symptoms of tinnitus, giddiness, pain, and excitement amounting sometimes to delirium), from meningitis, or congestion of the brain. More especially may the practitioner imagine the case to be one of cerebral origin, if the deafness, which may have occurred only on one side, have escaped his notice, and the local pain become so extended as to be no longer recognisable as proceeding from the ear. I believe that these formidable, but by no means uncommon, symptoms, are chiefly owing to pressure upon the fenestral openings into the labyrinth consequent on congestive swelling of the tympanic mucous membrane, or a large collection of mucus within the cavity. There is also an anatomical reason why these peculiar attacks of vertigo, and symptoms indicative of meningeal irritation should be so frequently contemporaneous with catarrhal inflammation of the tympanum and mastoid cells, and with the obstruction of the conduit, the Eustachian tube. You will recollect the arterial connexion which exists between the vessels of the dura mater and the cavity of the tympanum, and that consecutive inflammation of the internal ear may have supervened.

It must be once more reiterated, that imperviousness of the Eustachian tube, by preventing the escape of mucus from the tympanum into the pharynx, will cause retention and accumulations of morbific deposits; and that these, by keeping up constant irritation, will produce a baneful effect upon the tissues and contents of the middle ear, and will also eormously impair the hearing, by giving rise to adhesions and attachments of the ossicles to each other and to the walls of the cavity, and finally causing changes at the fenestræ of the labyrinth. Consequently experience teaches us that every person who has once suffered from severe catarrh of the ear, will be for some time constantly liable to a recurrence of the attack, and to relapses of deafness, from the presence of such irritating, clogging deposits, and of "membranous bands," the products of inflammatory action. In proportion to the number of them remaining after the attack, and the extent of the adhesions limiting the mobility of the ossicles and the vibrating membranes, will be the probability of recurring disease and the degree of deafness. Treatment, therefore, to be successful, must greatly depend upon our ability to prevent permanent thickening of the mucous membranes, and the formation of deposits which favour adhesions between the various parts within the tympanic cavity or Eustachian tube.

The *prognosis* of acute aural catarrh is, on the

whole, favourable when, 1st, it does not run on to suppuration; 2nd, if in the progress of the case, the membrana tympani does not become largely perforated. And 3rd, if it does not eventuate in any of the adhesive processes just enumerated, when secondary changes in the fenestræ or in the labyrinth have also occurred.

A certain proportion of these cases do unfortunately result in some deeper-seated or permanent affection, but these will be referred to hereafter when I come to speak of the *chronic* and *purulent* forms of aural catarrh. The number of these relapses, and bad consequences, would undoubtedly be much lessened were practitioners more generally able to recognise the incipient aural disorder; for it may be asserted with confidence that such affections of mucous membranes are quite as amenable to early treatment, and are as rapidly recovered from as other diseases, when they are energetically and properly attacked by treatment.

TREATMENT.

THE treatment of acute aural catarrh consists, in the first place, in the local abstraction of blood by means of leeches. Seldom will you have occasion to employ venesection or cupping behind the ears, as recommended by some authors. The employment of leeches is not only advisable because it is the most effectual and easiest mode of abstracting blood, but because they can be applied nearest to the seat of disease, and repeated as often as is necessary. From three to six should be made to attach themselves, some just in front of the tragus, and some below the lobe of the auricle. These are the points where they can be applied with the greatest advantage. If the mastoid process be the seat of inflammatory action, and especially if it be tender to the touch, they should likewise be affixed there, close to the angle between it and the auricle. It is sometimes desirable to abstract blood from the orifice of the meatus, particularly when the membrana tympani is actively inflamed; in which case the auditory canal must first be filled with cotton wool, to prevent the blood from falling to the bottom of the passage, coagulating there, and encrusting the surface of the tympanal membrane. From neglect of this precaution, instances have been known of leeches attaching themselves too far inwards, and causing extreme agony. I always mark the places where I wish them applied, with spots of ink, so as to avoid their being scattered about and employed uselessly, if not detrimentally.

The relief afforded by judicious leeching in acute catarrhal inflammation of the membrana tympani is often immediate, and perhaps more marked than in any other painful affection with which I am acquainted. The application may be repeated if the pain should not abate, or should it recur with severity. The bleeding should be encouraged by hot fomentations, stupes, poultices, or spongio-piline soaked in hot water. The moist heat of steaming, poulticing, &c. (with which a variety of sedatives may be combined), will be particularly grateful to the feelings of the patient. The vapour of very hot water medicated with opium, belladonna, conium, henbane, camphor, &c., directed to the fauces, gives comfort; or atomized fluids, composed with the same medicaments, in the form of spray, may be used in all inflammatory affections of the fauces, Eustachian tube, and cavity of the tympanum. Filling the external meatus hourly with water as hot as can be borne, and allowing it to remain for five or ten minutes, is an excellent way of applying both warmth and moisture as near to the seat of mischief as possible. The outer ear and contiguous parts may be unremittingly poulticed with "spongio-piline" soaked in medicated hot water, or with what is still better and more convenient, the new "epithema." This "ever-ready poultice and fomenting pad" is formed of flat pieces of sponge sewn up in fine linen, and a soft external waterproof covering. It retains the water, whether medicated or otherwise, and keeps it warmer for a longer time and without dripping than any other form of poultice known to me. It is besides very light, and can be fitted over the ears more closely than a linseed meal or bread poultice. Those who have suffered from the complaint now referred to, which becomes agonizing when the external surface of the membrana tympani is involved in the catarrhal inflammation, will testify to the almost magical effect of thorough and careful leeching, supplemented by hot douches and incessant poulticing, in relieving it. Cases of this character have been brought to me after days and days of treatment by blisters, stimulating drops and oils poured into the ear, without the slightest beneficial results, but which the use of a few leeches placed in front of the tragus succeeded in mitigating almost instantaneously. Leeching and hot water applications in some shape or other are therefore *the* local remedies on which you must rely.

I cannot advise you to give emetics or sternutatories at the beginning of these acute cases where exudations and softening of structure rapidly occur, by which the membrana tympani is rendered more liable to rupture. Even the *attempt* to inflate the tympanum is from the actively inflamed state of fauces, palatine arches, tonsils, &c., an effort producing pain; but an accomplished inflation (if the Eustachian tube has not already become entirely closed), is more acutely painful. Any agents, therefore, which cause sneezing or vomiting, are, though given with a view of getting rid of the too abundant mucous secretion or engorgement, inappropriate, because of a risk of perforation in the drum-head, which is (speaking comparatively) a rare consequence of the ailment we are now treating of. When there is much febrile disturbance accompanying the acute stage of this aural disease, mild aperients and diaphoretics are clearly indicated. They may be administered with morphia, hyoscyamus, conium, or any of the sedatives and narcotics which are the most suitable to and best borne by the patient—for the purpose of inducing sleep and lulling pain. The new drug (hydrate of chloral) will perhaps, from its tranquillizing hypnotic powers, supersede in a great measure the use of opium, being free from several of the deleterious properties of the latter.

It is usually recommended by authors, to have recourse "at once" to the use of mercury in some form or other (generally as calomel, with opium), and this, not only if the pain and deafness remain unrelieved, but even if the redness and vascularity of the external tympanal membrane are persistent. Inasmuch as within the last few years most of the acute ophthalmic and other disorders are treated successfully without the free use of calomel, or mercury in any form, and as the affection of the ear now under consideration may be compared (not inaptly) with those of analogous parts in the eye, we should endeavour to treat aural diseases in like manner, and on like principles. I am confident that you will not be disappointed in arriving at equally successful results by abstaining from the use of mercury. Recollect that the constitutional peculiarities commonly met with in patients who are the subjects of catarrh of the middle ear, are such as totally unfit them to endure the so-called " strictly antiphlogistic" measures recommended in almost all works on ear diseases. We must be careful not to reduce our patient's vital powers ; for if we do so, effusions of lymph, serum or mucus into the tympanic spaces will be probably increased ; and if by loss of structure a perforation of the membrana tympani take place, there will be a deficiency of reparative power to fill up the breach and restore the functional energy of the membrane. A "collapse" of the latter, with more or less disintegration of the ossicular chain will consequently occur. Weakly children and strumous young persons cannot bear well the frequent dosing with calomel, or even "grey powder," the usual panacea, according to popular belief, in all inflammatory complaints. A patient suffering from catarrhal disease of the ear is commonly disordered in general health; especially are the digestive functions disturbed. These will require appropriate treatment by medicines. If the pain in the ears, &c., ceases so as to allow of sleep without the aid of narcotics, they ought to be laid aside as soon as possible; for they almost always take away the appetite for food; and the patient is now in need of nourishment to repair breaches in structure, and to hasten the absorption of effused fluids—the products of inflammation. Lastly, with regard to general treatment, I will only observe that the more rational practice of the present day has effected a vast reform in the management of inflammatory diseases, whether situated externally or internally; and I cannot help expressing the hope that you, when in due time properly qualified practitioners, will make use of the same common sense and reasoning in the treatment of ear-diseases, which you will be called upon to apply to those of other organs of the body. An almost frightful system of general depletion, with severe and long-continued blisterings has hitherto prevailed because recommended in former works of high authority. You are compelled to adopt *local* depletory measures, for the purpose of relieving engorgements of the blood-vessels and disburdening the turgid capillaries that supply the mucous membrane of the middle ear, which is also the periosteum. Until this be done, the tormenting and dreadful **pain** will continue. Every one who has suffered from ear-ache (and few escape) has had proof of the intolerable nature of this pain. Generally speaking, however, I find that its *cause* is not rightly understood or interpreted, else we should not so commonly hear of the treatment being by rubefacients, blisters, chloroform, sedative liniments, and the like. It is true that pain in mucous membranes is not a common phenomenon, for their texture enables them to expand and dilate, so that they escape much tension and pressure; but the membrane lining the tympanum and its partition walls is altogether exceptional in structure, partaking quite as much of the character of a *serous* as of that of a *mucous* membrane. Consequently the signs as well as the results which belong to inflammation of both these tissues, will be usually manifested in an acute catarrhal affection of the middle ear. Every catarrh of the tympanum may be truly said to be a "periostitis" and we know how painful at the time, and how serious in its consequences, is an inflammatory condition of the investing and nutrient membrane of a bone. So that, in treating inflammations of these modified, or more properly, compound tissues, and adopting measures to prevent their oftentimes destructive sequelæ, we must be careful to bear in mind the close affinity which the so-called mucous membrane of the tympanic cavity bears to the proper serous membranes in other parts of the system. In this way, we shall understand how effusions of various fluids (such as serum, mucus, pus, or coagulable lymph which so frequently agglutinates contiguous surfaces, the formation of membranous bands, deposits, ulceration leading to perforation of the membrana tympani—persistent offensive otorrhœa, and lastly, caries of the temporal bone, with perhaps fatal intra-cranial disease), may, one or all, result from an aural catarrh affecting the lining membrane of the tympanum and mastoid cells.

The free application of leeches will mitigate the intensity of the pain by unloading the full and dilated capillary vessels which press upon the nerve filaments that ramify on the surface of bone (especially the promontory) and on the periphery of the membrana tympani. At the same time the probability of effusion will be lessened, and the desired event or termination of the inflammatory attack, viz., its subsidence or *resolution*, will be hastened. In the most favourable cases the disease will go no farther; the inflammation begins to recede, and there is very little or no appreciable effusion; or, should such have occurred to a slight extent, it is re-absorbed, and the parts return in most respects to their former condition and integrity.

Counter-irritation by blisters is, generally speaking, too much relied upon, or applied at too early a period in the disease, but as the latter advances, iodine painted on the mastoid process will be found useful.

The catarrhal symptoms in the throat must not be overlooked. In addition to the fomentations, steaming, &c., previously had recourse to, gargles, composed of opium and belladonna, will be beneficial as soon as the requisite movements of the palate and fauces can be borne. Borax, alum, cresote, iodine, &c., and astringents, may be employed at a later period.

By these means the object first aimed at, viz. *to subdue the urgent inflammatory symptoms*, will have been attained; but further measures are in most cases needful to prevent the *recurrence* of the ailment we are now considering, for it is the repetition of these attacks which is so frequently the cause of obstinate or confirmed deafness. One of the most immediate consequences of an aural catarrh is closure of the Eustachian tube, from the participation of the latter in the throat affection. As this imperviousness is only temporary, and will disappear on the subsidence of the acute faucial inflammation, I maintain that it is wholly unnecessary to resort to mechanical means for re-opening the tube at an early period, or until the patient can swallow without pain. On no account should the catheter be introduced while the severe inflammatory process continues, while the membrana tympani is greatly injected with blood, or when there is intense pain in the ear. When the pain and febrile symptoms have almost entirely disappeared, if the deafness be persistent and considerable; if the noises in the ear, such as blowing, singing, knocking, &c., &c., are troublesome; and lastly, if the patient be unable to inflate the drum, or ignorant how to do so, by the ordinary Valsalva process,—you may infer that the Eustachian tube continues obstructed, and should employ the Politzer method for re-opening it, without further delay.*

The obstruction being of recent origin is seldom

---

* This method has been thoroughly described and illustrated at pages 77 to 81.

great, and when the bag is compressed, air will probably pass freely up the Eustachian tube, and will be heard by the surgeon to enter the drum, with such sounds as indicate, with tolerable accuracy, the condition of that cavity.* The effect of inflation is sometimes surprising to the patient, and immediately beneficial to hearing. It generally brings about a restoration of the lost function, if the closure of the tube has been the most urgent abnormal condition. The little operation should be repeated daily, until the patient is able to press air into the tympanum by unassisted efforts, and the full degree of natural hearing is attained.

Great numbers of cases illustrative of the foregoing remarks and descriptions, might be adduced; but I am desirous to avoid what I cannot help regarding as a very common mistake in professional treatises, the multiplication of cases, often with scarcely any distinctive points, and tending at last rather to the bewilderment of the student than to his instruction. Two, however, are subjoined, being as nearly typical as possible, of the acute form of aural catarrh, ending in recovery.

CASE 1.—*Acute catarrhal inflammation of the mucous membrane of the middle ear. Eustachian tube obstructed by extension of disease from the throat. Cure.*

Madame ——, a justly distinguished vocalist, and a member of the Royal Society of Musicians, con-

---

* For a description of these sounds, during inflation, see page 93.

sulted me, as the aural surgeon to the society, in August, 1869. This lady had twenty years previously suffered so severely from a throat affection, attended by loss of voice, that she had resolved to relinquish her profession. In that condition and frame of mind she went to Dr. Yearsley, who, finding the right tonsil greatly enlarged, and the mucous membrane in its vicinity congested, removed the gland. The case is (by permission of the patient) recorded in Dr. Yearsley's book on "Throat Ailments," with the patient's name given. He says "All thickening was removed by the knife; and from that day she improved in health; the throat assumed a healthy appearance, the voice regained its power and improved in quality—in the latter respect to such a degree that its equal has not been met with in the opinion of many first-rate judges."

When the lady consulted me her face was tied up with a flannel bandage. She was led into the room looking pale and exhausted, and could scarcely walk. She had taken a violent cold after a debilitating illness. She complained of sharp, shooting pains, which seemed to be not only in the right ear, but to extend over the whole of that side of the head, the mastoid process, and down the jaw and neck. Pulse small and rapid; noises of a singing, hissing, beating kind in the ear, and general confusion in the head. Fever, and much constitutional disturbance, the latter aggravated by fear lest she should be unable to fulfil important engagements in Scotland a fortnight hence. The pain and soreness of the throat were increased by movement of the jaws, and consequently the mouth could not be opened sufficiently to allow of my thoroughly examining the fauces. A general redness of these parts was however ascertained to exist. The dermoid layer of the external auditory canal was involved in the inflammatory process (by extension from within the tympanic cavity), and the canal so constricted as to make a complete inspection of the membrana tympani impossible.* Any attempt at inflation by the Valsalva method caused suffering, and the permeability or otherwise of the Eustachian tube could not be well ascertained. The tick of my watch (ordinarily heard about thirty feet) could only be distinguished two inches from the right ear, and the vibrations of the tuning-fork when applied to the vertex were perceived only on the right side. I ordered the immediate application of leeches around the ear, and notwithstanding the general debility, directed the bleeding to be encouraged by fomentations and spongio-piline poulticing; hot water to be frequently poured into the ear, and warmed drops of liq. morphiæ hydroch. instilled into it. On the following day all the symptoms, subjective as well as objective, had improved, but I did not fatigue the patient by submitting the ear or throat to more minute examination. Two days afterwards I obtained a view of the membrana tympani: there was then no intense redness of surface, but its circumference was severely injected, and the whole mem-

---

* This, as you will observe, confirms the proposition advanced in the first of these lectures, that we must not limit *catarrh* to the mucous membrane of the middle ear only.

brane opaque, lustreless, and too concave. The manubrium of the malleus seen, accompanied by two or three distended blood-vessels. The throat less inflamed and painful, but still much congested. No power to inflate the drum, though the attempt occasioned no suffering. Hearing distance for the watch about the same as before, two inches. The fourth day all the symptoms had somewhat amended, especially was the pain abated by the leeching, &c. Three days afterwards, feeling assured that the persistent deafness was referable to the condition of the tympanum, and obstruction of the Eustachian tube, I employed my modified Politzer bag, and easily inflated the drum. The air passed freely into its cavity, but with a rather moist sound showing that some exudation had taken place. A trial of the hearing distance, now repeated, proved a wonderful increase, the watch, before distinguishable a few inches, was now heard many feet off. The noises were gone. The air-douche was used once or twice during the week. Improvement was rapid, and both local and constitutional inflammatory symptoms subsided. A chloroform liniment was substituted for the fermentations. A liberal diet and ferruginous tonics completed the cure of the disease. There was perfect restoration of hearing within the fortnight, when the patient took her departure to fulfil her engagements in Scotland with entire success.

CASE 2.—*Acute catarrh of the middle ear.—Regular course. Recovery.*

The following is extracted from Dr. Politzer's admirable work,* for it cannot but be of advantage to you, as students of aural surgery, to have the clinical experience of so excellent a surgeon and acute an observer, in addition to that of your own teacher :—

"Mr. A., student of medicine, in the month of June of the present year, shortly after a cold bath, experienced a severe pain in the right ear, with which a loud ringing was soon associated. An examination made on the following day revealed an uniform pink injection of the external meatus, especially in its osseous portion, the redness being more intense at its junction with the membrana tympani. This was marked at the anterior upper quadrant, near the short process, which, as a yellowish-white tubercle, was in strong contrast to its dark-red surrounding. Along the manubrium extended a dark-red bundle of vessels, so strongly developed as to make the handle itself invisible. Near the periphery of the membrane a circular wreath of vessels could be seen, from which small, slightly serpentine branches extended to the centre, to anastomose with the vessels of the manubrium. The remaining portions of the membrane lying between the injected parts, were dirty grey or lead coloured, and dotted with serous exudation. The light spot was dimmed and scarcely visible. The hearing distance was not much affected.

"On the Membrana Tympani in Health and Disease."

On account of the continued severe pain, five leeches were applied close in front of the tragus, and a gargle ordered of tinct. opii. ℨj., aq. ℥iv., with a little sugar. On the following day the pain had entirely left the right side, but had attacked the hitherto unaffected left ear, with the same severity and just the same appearances of the membrane as were upon the other side. Five leeches were now also applied upon the left side, and upon their falling off the pain immediately abated. On the third day of the attack, the pain on both sides had disappeared, but the ringing in both ears continued, and the hearing distance for the watch had fallen to about three or four inches on either side (the mean normal distance being twelve feet); for conversation it had fallen to somewhat more than six feet, and the condition of the membrane was nearly the same as on the previous day. After making both Eustachian tubes pervious simultaneously by means of the air douche. according to the method devised by me, the hearing distance for conversation immediately rose on the right side to twenty-one feet, and to forty-two on the left; the ringing was less, and the patient felt in all respects much relieved. On the fourth day of the attack the condition of the membranes was the same, except that they did not appear so moist. The hearing distance had sunk again to twelve feet, probably on account of the re-accumulation of mucus in the cavity of the tympanum, but rose to the height of yesterday after the employment of the air douche. Upon the fifth day a considerable abatement of the injection of the external meatus, and of the peripheral wreath of vessels was already apparent. Only the vessels of the malleus were still injected, and several small branches extended from the centre towards the circumference, which were sharply defined upon the dull, greenish-yellow membranes; the lustre of the membranes was entirely wanting. The hearing distance had decreased but little since the day before, and rose to fifty-four feet for conversation on the right side; on the left to forty-eight, and to three feet for the watch. During the three days following, the condition of the membrane as well as the hearing distance, remained the same. Upon the ninth day of the disease there was no trace of injection in the meatus, and upon the membrane only a pale red bundle of vessels could be seen along the manubrium. The cone of light was present, although dull and somewhat faded, and the membrane was of a dull gray. The hearing distance was nearly normal, and the ringing had entirely ceased. The air douche was continued daily. Upon the thirteenth day all the morbid appearances of the membrane had gone. The lustre and curvature, as well as the hearing distance, were perfectly normal. I had an opportunity of verifying this again by an examination some months later."

LECTURE IX.

SIMPLE AURAL CATARRH—(*continued*).

GENTLEMEN,—In the two last lectures I have noticed the course, principal symptoms, diagnosis, and treatment of the simple form of acute catarrh, when it terminates in *resolution*, or the restoration of parts to a healthy condition, and the faculty of hearing is duly regained.

In the following lecture I shall enter upon the consideration of cases where this form of aural inflammation does not run so favourable a course, but results, notwithstanding our treatment, in effusions of various kinds, which by accumulation greatly impair the hearing. This condition of the middle ear, if not recognised and specially treated by the surgeon, will be found to constitute the most frequent and subsequently irremediable cause of deafness.

Chronic catarrhal inflammation of the MUCOUS MEM-
BRANE OF THE TYMPANUM.

The *acute* kind of inflammation we have just been investigating may likewise be said to be characterized by severity of symptoms, with violence and rapidity of progress; whereas to the *chronic* kind (from *kronos*, time), with which we have now to deal, belong the opposite attributes of mildness, slowness, and long duration. We may also have every gradation of an inflammatory affection in the middle ear, beginning with that form in which we hardly can recognize the existence of increased action (*congestion*), and afterwards extending to such a violent degree as to interrupt or pervert all the ordinary functions, and even to alter or destroy the structure of the ear. It is necessary to point out in a general way one or two of these diversities; although the shades of difference included between the two extremes of acute and chronic inflammation are very numerous and indefinite.

Thus, we may consider acute catarrhal inflammation of the middle ear to be a violent disturbance, very painful and alarming, but it cannot last long; it will, like a fire, soon burn itself out. If not cut short by proper treatment it may end in ulceration or suppuration, or perhaps by extension to cerebral disease, or caries of some part of the temporal bone. In marked contrast to this stands the slow, languid, and indolent catarrhal inflammation belonging to the chronic form of the ailment, although this may perhaps in a considerable proportion of cases be only a continuation, with all the symptoms milder, from the acute stage. Also, in the chronic form of inflammation, the vascular distension and general disturbance are not nearly so great; and the redness of the mucous membranes which are visible is accordingly very much less, often scarcely perceptible on the faucial termination of the middle ear; while the pain is slight, or altogether absent. From this latter circumstance the patient may sometimes be quite unaware of the existence of very serious disease that has already been some time proceeding. Therefore, although in many respects chronic catarrh may appear less distressing and urgent than the acute form, we have become informed by post-mortem examinations that the pathological changes it produces in the ear are of a most serious character. These are interstitial deposits and thickenings, with such an amount of consequent structural alteration in the tympanic cavity and membranes as to gradually impair or totally destroy their varied and important functions, causing, it is needless to say, more or less severe deafness.

The question seems now to arise What are, in detail, the morbid or structural changes which we observe to take place in the cavity of the tympanum? where are they situated? and which are the important parts of the whole drum cavity most implicated in chronic catarrh when the deafness is of a high degree? In other words, what is the *pathology* of chronic catarrh of the tympanum? I need not tell you that it is only by carefully examining the parts of the ear in the dead subject, that we can learn the true nature of the varying morbid changes which occur during life.

In the first place we find that chronic catarrh of the tympanum results from an inflammatory state of the throat, which may have been either acute or subacute; and in these cases the mucous membrane of the cavity undergoes the process of hypertrophy. This thin and delicate investing membrane, naturally so fine as to require the sense of touch in aid of sight, even to determine its presence, may become so vascular, congested, and thickened as nearly to fill the whole cavity. From being "like a piece of the finest tissue paper," says Toynbee, "it becomes more like velvet." Instead of secreting just so much mucous as suffices to lubricate its surface, it pours forth viscid secretion in such excessive quantity as to occupy all the space remaining vacant in the cavity. In these cases, the Eustachian tube having been implicated in the catarrhal process, this fluid cannot escape naturally; and should it not become absorbed, it may, by pressure upon the membrana tympani, produce an orifice there, and run out into the external meatus. This latter symptom, or, more correctly, result, constitutes so frequent and important an ailment, taking its origin from more than one serious aural disease, that it will require special and detailed notice hereafter under its name of *otorrhœa*, or discharge from the ear.

Generally, in the cases now under consideration the whole of the membrane investing the tympanic walls is involved, and neither the thickening nor vascularity is decidedly greater in one part than another. In other instances, where there is a too abundant mucous secretion within the cavity, a thinner fluid may co-exist with a thick and tenacious one, the former occasionally disappearing in the mastoid cells, or escaping through the Eustachian tube when permeable, while the latter hangs about the ossicula auditûs, greatly impeding their mobility. When these mucous exudations become dry, dense, and stiff, they obviously may so interrupt the vibratile movement of the little bones, and of all the finely arranged membranes within the drum, and perhaps cover the so-called membrana secondaria at the round fenestral opening, as to act most disastrously upon the hearing powers. Thus much may suffice in reference to *exudations*.

Secondly, in pursuing our examination of pathological changes, we shall discover that the lining membrane itself sometimes becomes denser, firmer, and unyielding. This condition has been named by Toynbee, *rigidity of the mucous membrane of the tympanum*, and he has written almost a whole chapter on the subject. Whether it be possible or not to distinguish during life between cases of this description—cases of dried-up mucous collections, and those of fibrinous deposits which constitute the frequent pathological appearance called *membranous bands*—is of no great practical importance. It is unnecessary to describe these varieties of condition more fully. They all impair the motion of the ossicles; they may unite the little bones to one or another or to the walls of the tympanum; they may fill up all the angles and spaces in the cavity, or even obliterate it altogether by connecting the membrana tympani to its inner or labyrinthine wall.

Mr. Toynbee has described 136 cases of what he called *anchylosis of the stapes to the fenestra ovalis*. In my opinion this expression (anchylosis, or bony union) requires modifying, because he has applied it to those cases in which he found thickening and rigidity of the mucous membrane, as well as a similar state of the ligaments, of the stapedio-vestibular articulation, binding down and unnaturally fixing the stapes to the circumference of the fenestra ovalis. By true anchylosis, or stiff joint, we should understand that a bony union has taken place between the base of the stirrup-bone and the margin of the oval opening in which it properly ought to play somewhat after the manner of a piston in a cylinder. Sometimes such a condition is marked by an expansion of the base with new ossific matter thrown out so as to connect it with the adjacent parts of the oval fenestra; and sometimes the bony material is thrown out from the circumference of the fenestra ovalis, more especially from that rim or plate which, though it always escapes the notice of anatomists, exists, nevertheless, as a preventive to the forcible propulsion inwards of the stapes upon the labyrinth fluid. The pathological condition which results from chronic catarrh, and has been thus confounded with anchylosis, is dependent more upon contracted, tense, thick, or shrunken tissue, which holds down tightly the normally movable little stapes than upon calcareous effusions, or upon exostosis. I term this result of chronic catarrh *impaction of the stapes*, in contradistinction to Toynbee's anchylosis. Such a distinction appears to me necessary, for in the former case treatment is sometimes effectual, while in the latter no benefit can possibly be expected from it.

You will remark that some of these changes can be *seen* only upon the dead subject, because the catarrhal processes of which they are the result, may have established themselves upon parts comparatively far removed from the external membrane, as for example, on the inner wall and on the two fenestrae. In examining a living patient we are however assisted to conjecture, with tolerable certainty, that these two structures (next in acoustic importance to the auditory nerve itself), are implicated in the disease now being described, by the usually very high degree of deafness, which cannot be referred to a "nervous" affection. Any symptoms which indicate that the labyrinth-openings are partaking in the tympanic mischief, require our special study.

Chronic catarrhal inflammation of the MEMBRANA TYMPANI, EUSTACHIAN TUBE, *and* MASTOID CELLS.

I think that, to avoid confusion if possible, it will be best to give forthwith the description of the varied morbid appearances which are to be observed on the surface of the membrana tympani when submitted (in the living) to ocular inspection. If we combine with this the *pathology* of the Eustachian tube and mastoid cells, we shall, aided also by the previous summary of *post-mortem* investigations in diseases of this class, have sufficiently completed this part of the subject, and may proceed thereafter to consider symptoms, course, and diagnosis.

Although the structure of the membrana tympani

is very seldom affected alone, or primarily, generally becoming so by secondary implication, its aspect furnishes very valuable assistance in indicating the condition of the middle ear, and therefore importantly aids our diagnosis. If we examine the meatus of a young or middle-aged person suffering from chronic catarrh which has supervened upon the acute form, its cuticular surface will generally look swollen and moist, and it will be more or less filled with desquamated epidermis. On this account the boundary between the meatus and the membrana tympani will be ill-defined, and the superficial area of the latter consequently appears smaller. In other cases, where the disease is of a long standing, or the patient past the middle of life, all vascularity of the outer passage will have disappeared; there will be no excessive secretion or discharges, but it will be dry or scaly, indurated, rather shining, and totally devoid of cerumen. This last condition is usually associated with a pearly whiteness of the membrana tympani, and a line or two of red vessels running down the malleus handle. In others again, there is visible a flocculent exudation from the cuticular layer of the meatus and membrana tympani; the latter looks granulated and abnormally flattened, and according to the amount of this exudation, will the individual portions of the membrane be more or less indistinct. If you keep well in mind the position of the various structures composing the membrane, you will have no difficulty in distinguishing opacities caused by thickening of the *external* layer (the dermoid or cuticular) from those of the *inner* layer (the mucous).

For example, if the handle of the malleus is clearly seen, it is a sure sign that all the layers of the tympanic membrane (the fibrous, or middle layer included) external to this little bone, are free from deposits; but a very considerable thickening of the mucous layer may exist without hindering our view of the manubrium and short process. So also, according to the amount of exudation in its dermoid or cuticular covering will be the *colour** of the membrana tympani. Sometimes it is orange tinted, or greyish-yellow, and unevenly spotted. In opacities which are caused by chronic catarrh of the mucous layer, the membrane is usually somewhat dull, like parchment; and if it has not completely lost its lustre, may be compared to ground glass, or glass breathed upon. Its colour is often bluish-white, or dirty grey, and in long-continued cases, it passes through all the changes intermediate from " whitish-grey to a pearl white, from a bluish to a yellowish-grey " (Von Tröltsch).

The membrane is too concave, and consequently the *cone of light*† is altered, becoming dimmed, striped, or reduced to a small point, and sometimes entirely absent. Blood vessels are only seen on the membrane when congestion is present, and this is not often the case, though two or three may generally be seen coursing down the malleus-handle whenever the patient attempts to inflate the tympanum by the Valsalva method. The manubrium is, as before

---

* For the varying colour in health and disease of the membrane see page 44.

† See page 45.

stated, almost always distinctly visible, because it lies in front of the opaque mucous membrane. Sometimes this bone is very prominent, standing out as if in relief. In an uncommon case (that of Ed. Morris, a patient here) I made you observe the bone, so perfectly defined that I, at first sight, imagined the posterior half of the membrane to be lost; but upon inflation being practised, this portion which was so thinned and collapsed as to have fallen inwards against the inner wall of the tympanic cavity, became blown outwards, so as to be restored to its natural position. This man's severe deafness, of twelve years' duration, has been by this means and by one or two subsequent repetitions of the inflating process, completely cured.

In other instances, and often, the malleus is strongly drawn inwards by thickening and subsequent retraction of the mucous membrane and the tendon of the tensor tympani muscle; so that it appears in perspective *greatly foreshortened*, in which case the processus brevis or small round tubercle at its upper extremity, becomes more prominent, and is in fact the only portion of the bone seen. The appearance is particularly indicative of long-continued occlusion of the Eustachian tube, when the whole membrana tympani presents a most marked *inward curvature*, in consequence of the air becoming absorbed withinside the drum, while the usual atmospheric pressure (of about 7 lbs.) is exerted upon it from without. But this excessive concavity is not unfrequently observed, even when the Eustachian tube is not occluded; and in such a case, if the membrane does not move outwards, or only portions of it do so, during the employment of the air douche by the Politzer bag, or through the catheter—we may infer that adhesions exist on its inner side, within the tympanum, tying it to the walls of the latter. The movements of the membrana tympani, and the various forms which it assumes, are well brought into view by using Siegle's pneumatic speculum.*

Calcareous and atheromatous deposits which occasionally occur in the course of chronic catarrh, are easily recognized on inspection of the membrana tympani. They are by no means unfrequent, and when their presence is attended with a very high degree of deafness, it may be surmised that similar earthy deposits are present on the membrana of the fenestra rotunda, and about the foot of the stapes. Usually they form on the membrana tympani crescent-shaped or semilunar opacities of a whitish-grey colour, quite distinctly separated from the surrounding tissues. Wilde has likened them to the *arcus senilis* of the cornea. Since attention has been

---

* This ingenious instrument consists of a speculum at one end of a small air-tight vulcanite box, and a lens magnifying slightly, at the other, through which the surgeon looks. A piece of elastic tubing communicates with the box, through which suction can be made. When this contrivance is tightly inserted in the meatus, and the air exhausted, the membrana tympani may be seen through the lens to move, either in part or wholly, and thus the adherent portions are identified. I have been several times informed that an instrument of this kind is made use of by an aged aurist to, as he states, " draw the wax" (!)

more particularly drawn to these calcareous masses, it has been observed that they are very various in form, extent, locality, and structure. Sometimes they involve all the layers of the membrana tympani, and sometimes may exist only in the inner mucous layer, while the outer surface is unchanged in brilliancy.

Another of the many diversified appearances of the membrana tympani which come to view, more especially when the air douche is employed, is what have been called by Politzer "*bleb-like prominences.*" These may arise from loss of substance in the mucous and fibrous layers; in which case the impelled air gets between the latter (the substantia propria) and the dermoid layer; or there is a hernial protrusion of the mucous layer between the two external ones. From my own observations I can confirm these statements (first made by Von Tröltsch), and must add a third mode by which these appearances may be presumed to arise, occurring more frequently, and which I have noticed in my sixth Lecture, page 96. They are there designated by me from their origin, *emphysema of the membrana tympani*. Lastly, might be specified several forms of *bulgings* or *protrusions*, which are the result of interstitial exudations or of accumulations of serum, mucus, pus, &c., in the tympanic cavity.* A detailed description of all

* See Hinton "on accumulation of mucus within the tympanum, and its treatment by incision of the membrana tympani" (*Guy's Hospital Reports*, 1869); but *some* of the pathological changes in the membrane there described, are, I believe, due to causes other than those to which they are attributed.

these pathological changes in the membrana tympani is perfectly impracticable, and would, if given, be comprehended by only an earnest and devoted student of aural surgery. It is, however, only through a competent knowledge of the anatomy of the *healthy membrana tympani and of its morbid appearances in the living*,* that we can ever hope to fully understand the course of a chronic catarrhal affection, and to form thereby such a correct diagnosis as shall lead to successful treatment and permanent cure.

* **An** outline of the anatomy and appearances in health was given in Lecture III., page 40.

LECTURE X.

SIMPLE AURAL CATARRH—*(continued)*.

GENTLEMEN,—I now pass on to consider the *Eustachian tube in chronic aural catarrh.** It cannot be too steadily borne in mind that any abnormal condition of this canal must have an effect upon the cavity of the tympanum, and upon its external membrane. To produce impairment of hearing it is not necessary that the mucous membrane lining the tympanic cavity should always participate in the morbid changes caused by chronic catarrh at the other extremity of the tube (viz., in the pharynx, fauces, nose, &c.), for you have seen scores of cases in which the deafness has been very considerable, and of long duration, cured rapidly and completely by a few puffs with the improved Politzer bag. In such instances we may be tolerably certain that no impor-

---

* An outline of the anatomy and functions of the Eustachian tube is given at page 63.

tant *structural* disease exists in the cavity of the drum, in its lining membrane, or in the membrana tympani. Were it otherwise, the mere inflation, by restoring air to the inside of the tympanum, could not have given such instantaneous relief to the hardness of hearing.

I have already said much upon the mechanism of the Eustachian tube, the course of symptoms, and its general pathology* in acute catarrhal inflammation; but I still feel it necessary to dwell further upon the proofs of intimate connection between the diseases of the throat and of the middle ear,—that is between the pharyngeal and aural terminations of the Eustachian tube.

The following is an extract from Sir William Wilde's "Aural Surgery"†:—"Were we to put implicit faith in the writings of authors, or to quote authorities upon diseases of the Eustachian tube, we should be led to believe that the affections of that

---

* Lecture VIII., page 113; and those who desire to make themselves acquainted with the most recent investigations and modern views as to the functions and anatomical characters and relations of the Eustachian tube, will find advantage in reading Dr. Jago's recondite essay, published in the *British and Foreign Medico-Chirurgical Review* for January and July, 1867, Toynbee's "Aural Surgery," page 188, and an article in the *Lancet*, January 16, 1869. In this last some novel ideas, supported by personal experiments, are enunciated, and afford, I hope, materials for reconciling those differences of opinion which have hitherto existed among physiologists on this most important subject.

† Wilde's "Aural Surgery," page 365.

portion of the middle ear are of common occurrence. With, however, the exception of, &c., &c. [exceptions specified] I think that disease of that portion of the auditory apparatus is neither a frequent nor a usual cause of deafness. *Furthermore, it remains to be proved that an impervious condition of this canal is, as generally supposed, a cause of deafness.*"

Such are the words of our usually most correct, accomplished, and unprejudiced living British writer on ear disease. It is evident that he must have overlooked the close anatomical, as well as functional, connection between throat and ear, and, as a matter of course also, insufficiently noted their pathology. Many other authors have also touched upon this subject more slightly than its importance demands; and for these reasons I feel justified in keeping your attention fixed upon it for a longer time than I otherwise should have done.

The structure of the Eustachian tube, its continuation from the pharynx, its development and functional laws, and the effects of treatment, all confirm the impression that a catarrh of its lining membrane must almost necessarily result from a similar affection in the throat. Our daily observation leads to this opinion, as well as the frequent testimony of intelligent and unprejudiced patients who, without any pressing questions, almost always tell us that their deafness has proceeded "from a cold," or what is tantamount to it, a catarrhal ailment of some kind or another, including scarlatina, measles, &c. Finally, it may be taken as an established fact, that the Eustachian tube will readily participate both in the congestive swellings and exudations resulting from chronic catarrhal inflammation of the faucial mucous membrane, and that mere continuity of surface and similarity in structure, will cause the same disease to extend up its whole length into the tympanal cavity, and produce those effects already described as most destructive to hearing. But the inflammatory process, especially in the chronic form, may, and often does, stop short in some portion of the Eustachian tube, precisely as happens with inflammations of other mucous canals terminating in cavities, but which need not here be specified. It can be proved that any physical cause or morbid process whatever, which obstructs or closes the Eustachian tube, may, *per se*, occasion deafness of a very high degree, such as totally to debar the patient from hearing conversation unless addressed to him in loud tones, or shouted into the ears. I could adduce numerous instances (one is cited at page 14) in which, upon restoring the patency of this auro-tympanal canal, the extreme deafness instantly vanished. How this deafness was produced by a continuation of diseased conditions into the Eustachian tube, practical experience and a sufficient number of *post-mortem* examinations have enabled us to define with certainty. In the first place, any considerable hyperæmia, swelling or thickening of the pharyngeal mucous membrane will extend into the faucial end of the tube, and gradually pass on to the upper portion where it is naturally very narrow, and will in consequence easily become closed, and when it is closed, all communication between the throat and ear is cut off. No supply of air now being able to enter the drum, that which was contained within it at the time of closure will by degrees become absorbed, while the atmospheric pressure on the outside of the drum-head, continuing in its usual force, will press the membrana tympani as well as the whole chain of ossicles, inwards. The effect of this is, as you have several times been informed, to impede or destroy vibration in these parts to such an extent that sounds are prevented from reaching the internal ear.*

This deafness to external sounds will assuredly continue as long as the tube remains impervious; even for years, as in the Case No. 3, hereafter to be detailed. In the second place, closure of the tube will confine mucus, pus, and all other secretions or exudations within the tympanic cavity, and thus bring about an abnormal condition of this part of the middle ear—a prolific source of subsequent disease, which if long continued, may terminate in confirmed and irremediable deafness. This course or sequence of aural disease, from continuity of mucous structure, is, I do not hesitate to declare, in my opinion, beyond all others in frequency and in importance. We shall, therefore, not expend our time without profit in considering most carefully the chronic catarrhal affections of the throat and nasopharyngeal spaces, which thus so severely and injuriously influence the Eustachian tube and tympanum. One of the most immediate and commonest causes

---

* See page 117.

of *obstruction of the Eustachian tube is thickened mucous membrane.* A general hypertrophied state of the lining membrane of the fauces, soft palate, and posterior nares, is observed most generally in children and young persons of an unhealthy or scrofulous constitution. Occasionally all the mucous surfaces which can be brought to view on looking into the throat are reddened, congested, and swollen; and if at the same time the tonsils are enlarged, the boundaries and borders of the palatine arches are so approximated as to narrow and almost close the entrance into the pharynx.

An enlargement of these glands (the tonsils), so great as to make them meet in the centre, much interferes with deglutition and respiration. Small or soft morsels of food only are swallowed, and the child snores in its sleep, breathes habitually with the mouth open, and articulates badly. Sometimes, besides touching on the median line, the tonsils encroach upwards, towards or beyond the fine border of the soft palate, and downwards so low as to be hidden from view until the base of the tongue is pressed down. The mucous membrane covering them is almost always thick, and seldom without redness and increased vascularity. The diseased tonsils present an uneven, pitted surface, sometimes ulcerated in spots, very commonly small points of thick tenacious secretion of a yellowish color and muco-purulent character, exude from it, marking the orifices of some of the ducts which lead into the substance of the gland-tissue. This vitiated sticky exudation collects especially during the night, and covers the fauces and upper walls of the pharynx; doubtless also getting into the Eustachian tubes and preventing the free renewal of air in the tympanum when their mouths are opened. In these cases, and in others where the enlarged glands project above the soft palate, *deafness* is usually complained of, but this symptom is only a concomitant, and is not caused by actual pressure of the tonsils upon the orifices of the Eustachian tubes so as to close them, as was formerly supposed by Dr. Yearsley. The deafness appears rather to depend upon the thickened condition of the mucous membrane which extending from the tonsils and fauces, runs up into and lines the tube. For this reason there is always danger of chronic catarrhal affections producing deafness, even when the tonsillary enlargement is not great. Although the tonsils do not encroach so high up as to press directly upon the orifices of the tubes, they may cause a displacement of them by pushing up the posterior arch of the palate and the base of the uvula, occasioning much difficulty in passing the catheter, or even preventing altogether its being introduced.

Among the causes of defective hearing is also the growth of "vegetations," which spring from the walls of the naso-pharyngeal cavity.* Sometimes they

---

* These "granulations" were described by Dr. Wagner in the *Archiv. für Ohrenheilkunde*, vi., page 318, 1865; and quite recently Dr. Meyer of Copenhagen, wrote an excellent paper on the "Pathology, Diagnosis, and Treatment of *Adenoid Vegetations* in the Naso-Pharyngeal Cavity," which was read at a meeting of the Medico-Chirurgical Society, and appears in vol. liii. of their *Transactions*, 1870. These occur in groups, and smaller harder ones are found on the inner surface of the Eustachian tubes themselves. They frequently accompany chronic tonsillary disease, and a thickened or swelled condition of the soft palate, greatly impairing the mobility of the latter, and materially interfering with speech. The patient's voice is truly said by Dr. Meyer to be "singularly wanting in resonance, and the usual consonants cannot be pronounced, exactly as in a common cold; patients thus affected being unable to pronounce the nasal sounds, '*m*' or '*n*,' will say, 'cobbod' instead of 'common,' 'doze' or 'lose' instead of 'nose,' 'sogg' for 'song,' and being likewise unable to breathe through the nose, they are compelled to keep the mouth continually open." These vegetations or granulations act injuriously by keeping the surrounding mucous tissues in a state of chronic swelling, and in a reddish, puffy, infiltrated condition, which favours hypersecretion in the Eustachian canals. This will gradually increase, and be propagated to the tympanum, there to set up catarrhal symptoms of a more acute description, which may lead to obliteration of its cavity, or to what is less disastrous—perforation of the membrana

"adenoid growths" appear to me to be the same "granulations" described by Dr. Wagner, and which we cannot fail to notice if the throat has been thoroughly examined by the laryngoscope and with the finger.

You need not quite acquiesce in Dr. Meyer's opinion, that "a deaf patient who breathes through the mouth and has a thin compressed nose is affected with vegetations in the nasopharyngeal cavity, and to confirm this we do not even require to notice the speech."

tympani. (I refer you to Case No. 4.) Lastly, after what has been said regarding the multifarious causes of Eustachian-tube obstruction, you will be prepared to comprehend that "*ulcerated sore throat*" (to use the familiar term which is so well understood by everybody), hypertrophy of the pharyngeal glands, tumours of the soft palate and fauces, &c.—in fact, any morbid or catarrhal affection whatever which interferes with the perfect action of the Eustachian tube or its muscles, may impair the hearing temporarily, severely, or permanently.

Congenital fissure or "*cleft palate*" is a cause of deafness by deranging the action of the palatine muscles (which are, as you know, also muscles of the Eustachian tube), and otherwise by thickening the continuous parts of the mucous membrane. The effects being catarrhal, I must notice them. Dieffenbach has shown that "most patients with cleft palate are hard of hearing."*

In addition to *mucous plugs and coverings* at the orifices of the tubes, nasal or pharyngeal *polypi*, large *ulcerated surfaces* (specific or otherwise), *membranous adhesions*, *stricture*, and other conditions more rarely occurring as causes of obstruction in the Eustachian tube, I must not omit to mention what is called *relaxation of the mucous membrane of the fauces*. I do not class this latter affection with those just described as thickenings, though the symptoms are not

* In the *Lancet* for March 6, 1869, I reported a case where deafness dependent upon cleft palate in a child seven years of age was cured by the use of the pneumatic speculum and by simple applications to the throat.

dissimilar, because it more generally occurs in adults and perons who are debilitated from overwork, and are subject to indigestion. It also is seen in those who are over-indulgent, and irregular in their habits.

The physical cause of the obstruction (as described by Toynbee) appears to be a relaxed condition of the mucous membrane covering the faucial orifice of the tube, so that its muscles are unable to separate the lips sufficiently to admit air. The exciting cause is often a cold. There is usually no history of a previous disease of the ears, the deafness coming on slowly, and gradually increasing until the patient is unable to hear ordinary conversation, and requires to be spoken to in a loud tone of voice.* It is remarked by Mr. Hinton that "those who smoke to excess are especially liable to deafness from this cause." My own observations, however, have hitherto failed me in tracing the infirmity to such a habit.

Middle-aged patients suffering in this way are generally weak and debilitated, looking pallid and out of tone. On examination, the mucous membrane of the fauces is less red than usual, or pale and flabby; it may be hanging in folds on each side the arches of the palate, and traversed by streaky blood-vessels or distended veins. The "*relaxed uvula*," in consequence of the palatine muscles having lost their contractile power, is too pendulous, sometimes hanging so low as to touch and rest upon the dorsum of the tongue. The term "relaxed" does not, how-

---

* "Diseases of the Ear," page 211, 212.

ever, always describe correctly the altered condition of the uvula, for it may be thickened and not elongated, but the latter is the more usual appearance. It may be œdematous, infiltrated with serum, and softer than natural, or obliquely placed between the tonsils and the arches of the palate. Occasionally, on attempting to pass the catheter, after failure of inflation by the Politzer bag to restore patency in the tubes, we come upon a projecting mass of relaxed mucous membrane which has arched the soft palate forwards, and gives a peculiar "doughy" feel at the end of the instrument. I have several times found resistance to the passage of the latter into the Eustachian tube from such folding of the membrane which seems to act like a valve at its faucial orifice. Such are instances of *relaxed* mucous membrane.

Chronic catarrhal inflammation of the MASTOID CELLS.

As it seems to be doubtful whether the mastoid cells are ever primarily affected, and without contemporaneous disease of the tympanic cavity, I have not noticed their pathology before when treating of the acute form of aural catarrh. That they become secondarily affected, by extension of inflammatory action from the tympanum, is a matter of every-day experience.

Perhaps the only exception which can be allowed in respect of an idiopathic affection, in which the tympanum is not involved (or does not appear to be involved), is when certain fatty, chalky, scrofulous or other deposits occur in the mastoid cells, and are not at the same time found in the tympanum. In otitis, or the purulent form of aural catarrh, the mastoid process seldom escapes implication; but this serious disease, leading occasionally to a fatal result, belongs more properly to a future part of our subject, and will be attended to later. We cannot, however, understand the pathology of the mastoid cells, nor even how far a chronic catarrh in them affects the hearing, until we have acquired a clear idea of their functions and physiological importance; for in this way only can we determine the amount of significance to be attached to any particular manifested disease in these parts. Mr. Toynbee pointed out, that at birth and during childhood, the mastoid process is in a rudimentary state, and its form is very different indeed from what it becomes later on in life. He also carefully demonstrated by dissection and *post-mortem* examinations of the temporal bones of children who had died of caries and abscess in the brain from the effects of catarrhal inflammation, that disease occurring in the mastoid process during childhood produces entirely different results from those which follow it when occurring at a later period. You must therefore make yourselves thoroughly acquainted with the peculiar conformation of these cells, and their anatomical relations to the tympanum, before entering upon the study of their diseases. By dissection with your own hands, and not from descriptions given by another person, can such knowledge be obtained.

With regard to *function*, the mastoid cells may be considered as an amplification of the tympanum, or an appendage to that cavity. It seems to be an accepted opinion that this porous, light, yet firm and stable cellular structure acts as a sort of reservoir for air to the drum, with which the cells freely communicate. Being placed immediately opposite to the entrance of the Eustachian tube, air easily passes into the mastoid cells, and is set in motion by acoustic vibrations impinging upon the membrana tympani. And when that membrana is extensively perforated, or lost through disease, the air contained in this amalgamated cavity is influenced greatly by the sonorous impulses from without. By bearing the last-mentioned circumstances in mind, you will understand how an excellent degree of hearing may be attained, even under such deficiency, when the little plug of cotton wool (Yearsley's artificial tympanum) is inserted against the ossicles, or sometimes against the stapes, if that be the only one remaining.

It has often happened, that when the mastoid cells have been filled up, by congestive swelling of their lining membrane, or by accumulations of mucus, pus, &c., and the air contained in them has thus become greatly diminished, the membrana tympani has suddenly burst during sneezing, coughing, or a violent blowing of the nose; and my opinion is, that were it not for this auxiliary arrangement of pouches into which air may be occasionally compressed, any sudden, loud, and unexpected sounds, such as gun-firing, the explosion of shells in warfare, or even a box on the ear, would cause rupture of the drum-head to be much more frequent than it is. In truth, it is quite an allowable " scientific use of the imagination " to state that if the tympanum (which you know is only the size of a small nutshell) were the sole air-containing cavity, few of us would be in possession of a sound unruptured drum-head.

SYMPTOMS *of chronic aural catarrh.*

The objective symptoms, or appearances of those parts of the middle ear liable to chronic catarrh, and which can be brought under inspection, having already been detailed, it only now remains for me to notice the subjective signs, or those experienced by the patient. *Deafness* is generally the most prominent one complained of; although it may be so trivial in degree that attention has not been seriously called to it until singing in the ears, or some other distressing sensation has become sufficiently urgent to impel the patient to seek advice. The defect in hearing has perhaps increased so gradually and imperceptibly that no particular time or cause can be assigned. Catarrhal affections of the middle ear so frequently have commenced in infancy or childhood that there is difficulty in getting *history*, and in reply to the question as to "duration of disease," we are often told, " I was deaf from infancy," " I never recollect hearing well," and the like. Again, if the patients' attention has not been drawn to their infirmity by pain, tinnitus, irritation of the ears, or any other unusual sensation, it is quite common for them and their friends to consider the case as one of " nervous deafness." Fortunately, however, this illusion is dispelled in a large proportion of these instances by our modern and more exact methods of examination. This slow and insidious advance of deafness indicates that the mischief is the more serious, probably thickening of the mucous membrane of the tympanic cavity, but the degree of deafness depends not so much upon the amount or extent of the thickening, exudations, or depositions, as upon their situation. For instance, we find that extensive morbid alterations, and even losses of structure in the membrana tympani and the mastoid cells may occur without much hardness of hearing; while on the other hand, very slight changes or deposits on the membrane covering the fenestræ and in the parts which conduct sound to the labyrinth, will impair hearing to such an extent as to cause almost total deafness. The *degree* of deafness will consequently, with other symptoms, indicate tolerably well the locality implicated.

Tinnitus, or singing in the ears, is one of the many symptoms, and the commonest subjective one, and often forms the principal complaint, when the deafness has slowly stolen on. Wilde,* describing these aural sensations, says:—

"Noises in the ears" is certainly one of the most distressing, as well as the most frequent symptom attendant upon affections of the organs of hearing. It is one common to almost all, and peculiar to none of the diseases of the ear. Like muscæ volitantes in the eye, it may exist as an isolated symptom, or it may be an attendant upon several aural diseases. It is often caused by cerebral disease; it is sometimes an accompaniment of derangement of the circulatory, digestive, or uterine organs; of congestion of the

---

* "Aural Surgery," pages 83 and 84.

SYMPTOMS OF—TINNITUS.

brain, hæmorrhage, hypochondria, hysteria, chlorosis, anæmia, typhus, influenza, or simple catarrh; of closure of the external meatus, obstruction of the Eustachian tube, and impaction of the auditory passage with wax; a foreign body, or even a hair resting on the tympanal membrane, as well as engorgement of the lining membrane, or mucous collections in the tympanic cavity, and also nervous deafness—these will all produce it. Furthermore, we may remove the original disease, give a healthy action to the affected organ, and restore its function, yet the noise will remain. So great is the discomfort which it gives that persons incurably deaf, and who are quite conscious of the impossibility of restoring their hearing will still apply to be relieved from this haunting and most annoying symptom. The peculiar characters of the *tinnitus* and the noises to which it is likened, are as variable as sound itself. I think the descriptions which patients give of the noise which they experience depend, to a certain degree, upon their fancy, their graphic powers of explanation, and not unfrequently upon their rank in life, or the position in which they have been placed, and the sounds with which they are most familiar—thus, persons from the country, or rural districts, draw their similitudes from the objects and noises by which they have been surrounded, as the falling and rushing of water, the singing of birds, buzzing of bees, and the waving or rustling of trees; while, on the other hand, persons living in towns, or in the vicinity of machinery or manufactures, say that they hear the rolling of carriages, hammerings, and the various noises caused by steam engines. Servants almost invariably add to their other complaints that they suffer from 'the ringing of bells' in their ears. The tidal sound, or that which we can produce by holding a conch-shell to the ear, is, however, most frequently complained of. Sometimes the tinnitus exists as an isolated symptom; but in several such cases I have remarked that sooner or later either aural or cerebral disease manifested itself. Removing the cause, and curing the deafness will often, but not always, relieve the patient of the noise."

With every line I have read to you I cordially agree; but my present object is to treat of this subjective aural symptom only so far as it relates to and depends upon a chronic catarrhal affection of the ear. Morbid anatomy, more accurate methods of examination, and a stricter observation of the physical signs accompanying this disease, have in a great measure cleared up the mystery surrounding the production of "noises in the ears or head." As chronic catarrh is the commonest form of deafness, so is tinnitus aurium the most frequent result or sign of it. I will briefly enumerate the abnormal conditions of the various parts of the middle ear which appear to be the usual exciting causes of this disturbing symptom. First, the state of the membrana tympani. In acute catarrh of this part it is always present (see page 111), and we likewise find tinnitus accompanying the more chronic form of catarrh when there is overtension, or the opposite condition of collapse, or whenever the membrana tympani is rendered invibratile. Pressure of wax,* and fluid or thick discharges on the outside, accumulations of serum, mucus, or other fluid, or semi-fluid collections, earthy or fatty deposits, and membranous bands on the inside, are causes of tinnitus. Even such hyperæmia and slight amount of hypersecretion of the mucous membrane of the tympanic cavity as occurs during an attack of influenza, common cold, or sore throat, or when there is increased circulation from indulgence in stimulants, taking quinine and strong ferruginous tonics, are all causes sufficient to give rise to tinnitus, and if, in addition, the Eustachian tube be closed, the noise in the ears is sure to be aggravated.† The supposition that the condition of the membrana tympani is the chief agent in the production of tinnitus aurium derives strong probability from the fact that when a perforation of it, or any large loss of its substance from ulceration or the suppurative process occurs, through which secretions, &c., from within the tympanum are freely discharged, tinnitus is scarcely ever present. The noises also cease when we incise or

---

* See a case of this kind, rather peculiar from its having been overlooked, related on page 11.

† A physician, eminent in a special department at one of our hospitals, applied the other day to me on account of this troublesome symptom, unaccompanied by much deafness. I ascertained that he was taking iron, with three and-a-half grain doses of quinine, and thus had produced congestion of the mucous membrane of the Eustachian tube (and probably also to a slight extent a similar state of the tympanic cavity), so as temporally to close the former. A puff or two with the Politzer bag cured him.

perforate it artificially, so that this operation is occasionally submitted to by patients for the purpose of obtaining relief from them.

Now, remembering what has been stated regarding the functions and diseases of the tympanic structure, and noting the foregoing instances, you will not have failed to perceive that tinnitus has depended upon some abnormal pressure upon the nervous expansion in the labyrinth. The membrana tympani presses the ossicula inwards, and therefore the base of the stapes upon the fluid where the auditory nerve is distributed; or it may be so rigid, tense, and unyielding, that the secretions within the drum press unduly upon the still more delicate membrane of the fenestra rotunda. Thickening and great tension of the lining tympanic membrane do the same thing. When, in a case of aural catarrh, tinnitus and deafness are simultaneous in their commencement, they will increase proportionately, and it is obvious that in this instance both must depend upon some alteration in the conducting apparatus, by which its acoustic properties have been interfered with. For example, a little film of mucus, spread over the inner side of the membrana tympani, is sufficient to alter the periodicity of its atmospheric vibrations, or even partially to quench them. Thus, deafness and tinnitus will co-exist here. But as soon as the removal or dispersion of the coating from the membrana tympani occurs (generally indicated by a pop or crack), both symptoms will together instantaneously vanish, perhaps never to return until the next attack of catarrh. We are amply warranted by facts like these in concluding that the membrana tympani is generally in some way or another concerned in causing tinnitus.

Sometimes distension of the blood-vessels in the tympanic cavity renders the circulation within them audible to the patient, who is sensible of a kind of "beating sound," which can be diminished, or altogether stopped by pressure upon the carotid artery. So resonant, in fact, is the cavity of the drum, that we find provision made against the occurrence within it of even such minute noise as the contraction of the fibres of the tiny ossicular muscles. These are shielded in bony canals, and only their slender tendons enter the drum, to become attached to the malleus and stapes.* Paralysis of the tensor tympani, or, as shown by Politzer, shortening of the tendon belonging to this muscle, may, whether accompanied or not by any other condition tending to

---

* Several years since, when treating (not in print) of the various functions of the tympanum, I endeavoured to demonstrate that the contractions of the stapedius and tensor tympani muscles would inevitably be heard, to the confusion of other sonorous vibrations, were their fleshy fibres not encased in such bony surroundings as the pyramid and the other distinct canal parallel with the Eustachian tube. More recently (in the *Medico-Chirurgical Review*, April, 1867), Dr. Jago has drawn attention to the same anatomical fact, and apparently attributes to it the same significance as I have done, when he remarks, "Hence there is a careful provision against the conveyance of the sounds of these contractions of muscles to the auditory path." Dr. Jago is, as far as I am aware, the only writer of eminence on acoustic physiology who has noticed the important bearing which such a mechanical arrangement must have upon the tympanic functions.

increase intra-auricular pressure, of course cause most troublesome tinnitus; so that the almost universal occurrence of this symptom in chronic catarrh may be expected.

Next in frequency to interference with the membrana tympani, arising from one or other of the pathological changes above referred to, I am inclined to place closure of the Eustachian tube as the most common cause of "singing in the ears." This also, on analysis, proves to be such, not directly, but in the manner following:—A closed tube necessitates a too great curvature inwards of the membrana tympani, and consequently an abnormal pressure upon the nervous expansion within the labyrinth, as has so repeatedly been explained.

Another characteristic symptom in chronic catarrh bearing some analogy to tinnitus, is a feeling of *heaviness and fulness* in the ears, as if they were "stopped up," which is increased by any cause producing interference with the aural or cerebral circulation. A hearty meal, drinking wine, violent or unwonted exercise, or any similar excitants and stimulants will sometimes occasion it, and so will mental or bodily fatigue, nervous depression, sedentary occupations, and too close application to study. These sensations, or some form of tinnitus, usually accompany every severe cold in the head, and they will sometimes linger after the high degree of deafness has passed away, although more commonly they increase or decrease, *pari passu*, with the hearing power. We must consider these cases as less hopeful of relief, or cure by treatment, than those in which the impairment of hearing is unattended by any of the subjective symptoms we are now discussing; because, when they persist, actual pathological changes are probably added to the hyperæmia of the tympanic mucous membrane.

Certain blowing, hissing, or pulsating sounds occasionally complained of, may be referred to some irregularity in the vascular system of the middle ear. Enlargement or even constriction of the capillaries in the tympanum would transmit a noise to the labyrinth,—so extremely resonant is the former portion of the organ; and I agree on this point with Dr. Jago, rather than with Mr. Hinton, who thinks "it may be held probable that any considerable amount of tinnitus seldom exists without somewhat of morbidly increased irritability of the auditory nerve."* The above-mentioned symptoms of fulness and pressure in the ears may increase to severe headache; great nervous irritability, vertigo, and even sickness may also occur. These appear strange ailments to be thus connected with a chronic aural catarrh, yet we not unfrequently find that a purely *local* treatment relieves the patient from them, especially when the air-douche is used. A feeling of lightness, quiet and freedom takes the place of noise, fulness and oppression.

Time will not permit me to go into this matter

---

* The pathology of this subject has, I would fain believe, made progress since Mr. Hinton, in his supplement to Toynbee's work (1867), wrote (page 463), "It would seem better that the causes of tinnitus should be held as yet a very open question."

so fully as I could wish. The significance of tinnitus, as almost invariably a concomitant of chronic aural catarrh, is certainly better understood at present than formerly, and, therefore, it ought now to be less empirically treated. The subject may perhaps be summarized by saying that those forms of tinnitus which relate most to sensation, may be considered as a kind of neuralgia of a reflex character, induced by contemporaneous affection of the nerves belonging to the tympanum and adjacent part, such as the trigeminus, glosso-pharyngeal, the otic ganglion, and sympathetic plexuses; while the severer symptoms of the same class, such as vomiting, and vertigo, &c., are to be associated with intra-auricular pressure (that is, pressure upon the labyrinth fluid, or on the auditory nerve-expansion itself), in a manner already explained.

The facts referred to in the foregoing pages, and the observations and deductions to be drawn from them, *must* be made thoroughly your own, if you would avoid the common error of imputing tinnitus, really arising from catarrhal inflammation of the middle ear, to a *primary* affection of the auditory nerve. Unless we are able to connect this most important, distressing, and undefinable symptom, with the discoverable morbid conditions in the ear itself, we shall never diminish the number of cases of "nervous deafness," so called,—certainly shall never reduce them to the proportion afforded by my own experience, that is, to about one per cent. of all aural diseases.

LECTURE XI.

CHRONIC AURAL CATARRH.—(*Continued*).

GENTLEMEN,—The appearances of the *throat*, in both acute and chronic aural catarrh, have been rather fully described when speaking of those parts of the middle ear most immediately connected with the naso-pharyngeal mucous membrane; (see pages 113–115, the Eustachian tubes), so that it is unnecessary to dwell longer upon them in this lecture.

The more closely we study the throat symptoms in all aural diseases, the more frequently shall we find facts demonstrating that pharyngeal affections, of whatever nature or origin, are most intimately associated with catarrhal inflammations of the ear. Accumulations of phlegm, or the opposite condition of unusual dryness of the throat and nose, the altered character of expectoration, the odour of the breath, a certain difficulty of swallowing, the tickling cough, the open mouth through which the patient habitually breathes, snoring, thickened speech, &c., &c., are all local symptoms which more or less declare themselves in a deafness which has been primarily caused by a throat affection, and then has continued. These are such common accompaniments of chronic aural catarrh, that it would be a waste of time to describe them more in detail. Though patients may oftentimes forget to specify that they have experienced any or all of them at some period during the course of their malady, yet upon being closely questioned, they will generally be able to refer the accession of their deafness to some such evidences of throat ailment. Recollect that the soft palate, fauces, nose, Eustachian tube, pharynx, cavity of the drum, and salivary glands, are very fully supplied with nerves, both motor and sensory, of varied origin and distribution; and that, therefore, when these parts are morbidly affected, other organs, though remote in their position, easily become involved. There are few portions of the human organism which are brought into such intimate and various connection with different nerve-tracks, as is the throat; and it follows as no matter of surprise that catarrhal disease of an organ so near to and directly continuous with the faucial mucous membrane as the middle ear, should sometimes remain long after the primary ailment has been forgotten.

The *pain* occurring in this form of aural catarrh is certainly not a prominent symptom; not indeed a frequent one; but when it continues for a long time, it is generally indicative of a more subacute stage in the disease. Usually felt when the patient is exposed to cold blasts of wind, or draughts, it is described as "gnawing" or "biting" in character, and will be only transient. If closure of the Eustachian tube have ensued rapidly from adhesive inflammation, and evidences exist of thickening and other structural changes, and of deposits in the tympanic cavity, severe pain will sometimes occur.

DIAGNOSIS *of chronic aural catarrh.*

I have now spoken to you of the varied appearances of the membrana tympani under chronic aural catarrh, of the symptoms (objective and subjective) and course of that complaint, as well as of the morbid conditions produced by it in the throat, Eustachian tube, and cavity of the tympanum. Also I somewhat minutely detailed, in the earlier part of these lectures, the results of inflating the middle ear by the Valsalva and Politzer methods, and by catheterism of the Eustachian tube (Lectures V. and VI., pp. 70 to 98). We have thus seen of what high value is the assistance of auscultation in the diagnosis of all chronic aural catarrhs. By its practice we become informed whether the Eustachian tube is freely or moderately penetrable by a stream of air, whether the same canal is swelled or constricted in calibre, and whether there is any unusual amount of mucus either in the tube or in the tympanic cavity. Subsequently, by inspection of the membrana tympani, and studying the mechanical effects produced upon it by the air-douche, together with other general observations, we shall probably have determined with some precision the nature of the alterations which have caused the impairment of hearing.

It is necessary, however, to guard carefully against inferring too much from any one mode of diagnosis. An inspection, however accurate, of the membrana tympani only, is not sufficient to enable us to judge decisively of the amount of deafness, nor of the nature of the morbid processes going on within the cavity,—although as a rule, it will furnish valuable information with regard to both. Neither does the employment of the air-douche in any of the ways specified, combined with a very correct estimate of the results of inflation, alone guide us to unerring conclusions as to the condition of the ossicles, the important fenestral membranes, or the membrana tympani. For example: an aural catarrh that has localised itself in the drum-cavity will cause little or no alteration in the Eustachian tube, and therefore would not be diagnosed by auscultation; yet in such an instance, the impairment of hearing is probably the result of a morbid state of the tympanum, which has been primarily induced by inflammation and afterwards closure of the Eustachian tube. The catarrh of the tube may have terminated in resolution, and its patency become restored, while the products of inflammation have remained in the drum and cannot escape.

The appearances of the membrana tympani during chronic catarrh affecting itself and the tympanum, have been so amply detailed in the 9th Lecture (pages 143 to 149) that I shall not again specify them. The sounds heard when we listen to them through the otoscope have also been particularised (see Lecture VI. under the heading " Effects of Catheterism," pages 92 to 98). But as in the diagnosis of *obstructed Eustachian tubes*, most essential help is afforded by the use of the inflating bag or catheter, and by inspection of the membrana tympani during and after the operation, I shall touch once more upon the most important points to be observed by the surgeon in making his examination. The real value to be attached to each particular symptom indicating obstruction, must be thought over and worked out by yourselves.

The *evidences* of an obstructed Eustachian tube are chiefly as follow :—a changed position and altered curvature of the membrana tympani. It is too concave, and looks sunken inwards, especially in the central portion. The malleus-handle appears on this account foreshortened, and lies nearly horizontal to the meatus, while its rounded tubercle or short process becomes very prominent—like the head of a large pin—near the upper circumference of the membrane. The membrana tympani being thus drawn inwards, and perhaps adherent to the promontory, the latter is seen shining through it. The vertical process of the second bone in the ossicular chain (the incus) on which the membrane in such a case often lies, is occasionally to be distinguished. From these causes, the triangular spot of light is altered, or may be altogether absent. Sometimes one half only of the membrane is depressed and indrawn, in which case it is generally the anterior segment. The colour and consistency are altered. It may look thinner, as in atrophy of its fibrous layer, and flabby, presenting the aspect called by Wilde "collapse;"

and if air can be forced through the Eustachian tube, to drive it from within outwards, or Siegle's apparatus be used to suck or draw it outwards, portions of it may be seen to move in this direction, and then to fall back again to the former position. If, however, thickening has resulted from the chronic catarrh, the signs of a previous closure of the tube are not well marked, because the membrane will be less moveable.

If only certain portions of the membrane are either thinned, thickened, or adherent, such alterations will of course become very distinctly visible under inflation or suction. The appearances indicative of accumulations and deposits within the cavity, which are the consequence of prolonged imperviousness of the tube, will be recognized in individual cases, and they have been mentioned before.

Pathological investigations have proved that in some very rare cases of decided catarrhal deafness, no abnormal changes whatever were visible on the membrana tympani. Such instances, however, are most exceptional, for whenever the membrane lining the cavity of the tympanum is much affected, the mucous layer which is continued upon the drum-head is in the same morbid condition, and this can (almost invariably) be recognized on inspection.

For a description of the diagnostic signs of various catarrhal processes, which are afforded us by listening through the otoscope (*auscultation*), I must refer you to a former lecture (page 93). The more the examiner's ear be practised in auscultation, the more valuable to him in diagnosis will be the sounds heard. Each case must be carefully analysed by itself, for the mechanical effects produced by catheterism and by the air-douche, are never precisely alike in any two instances;—so much depending upon the nature and degree of closure of the tube, the amount and character of fluid secretion in it or in the cavitas tympani, the firmness and extent of the abnormal adhesions, and so on. Other valuable aids in the diagnosis are available: such as the history of the case, the patient's statements; the origin of the disease; the kind, duration, and degree of deafness; the external influences, if any, which affect hearing; whether it is better in a noise; and whether high or low sounds are the more easily distinguished. The metronome is a good instrument for testing this. Examine the throat, nose, &c., and ascertain if sudden changes in the hearing power are apt to occur; for if they do, it may be assumed that the Eustachian tube is in some way more involved than the cavitas tympani. Some patients will state that they always hear better in dry weather, and that when the atmosphere is charged with moisture, their hearing power diminishes. These hints should be taken advantage of; for they lead to the inference that the mucous membrane lining the tubes is in a thickened and congested state, and that the slight additional swelling or secretion which during wet or foggy weather always takes place in membranes accessible to the air, is sufficient to bring the inner walls of the tubes together, thus rendering them less pervious to air, or sometimes entirely obstructing its passage into the drum. Hence you find that some improvement is occasionally obtained by such patients restoring to the Valsalva method of inflating the drum by swallowing with the nostrils closed. Again, when the deafness is not variable, nor the hearing improved by restoring air to the interior of the tympanum through any of these means, it may be assumed that the mischief lies in that cavity. Also, if the hearing power be very seriously impaired, although the auditory nerve is ascertained to remain unaffected, it may be safely diagnosed that structural alterations have occurred at the two fenestræ which lead into the labyrinth.

At this point, I must ask your attention to what is called by Von Tröltsch "a great gap in our knowledge." He says "We have no means, as yet, of determining on the living subject, the particular seat of the morbid changes, that is, any more exactly than has just been indicated."* He also quotes Politzer on this point, who says "We have no way of determining whether a given impairment of function depends on an adhesion of the head of the malleus to the upper wall of the tympanum, for example, or upon a diminished mobility of the stapes in the fenestra ovalis." Von Tröltsch adds "Perhaps an examination with the tuning fork may lead to some conclusions on this point." Another author,† writing so lately as this year, says, in reference to anchylosis of the stapes, "I do not know any symptoms by which it can certainly be distinguished during life."

* Von Tröltsch "On the Ear," page 334. Dr. St. John Roosa's translation.
† Hinton "On Diseases of the Ear," in Holmes's Surgery, page 310. Edition 1870.

It would take too much time were I now to enter at any considerable length into my reasons for the belief that diagnostic signs are undoubtedly often manifest, by which we may with tolerable certainty determine the exact seat of the catarrhal disease, when located in the tympanic cavity or in some portion of its contents. I will nevertheless advert while on this subject, to some observations made by me in an article published in the *Lancet* of January 2nd, 1869. I quote from it as follows:—" We assuredly may diagnose rigidity of the ossicles or anchylosis of the stapes, by careful attention to the symptoms accompanying their progressive stages, by the history of the case, and by observing the degree of loss which the patient has sustained over what I term the power of adjusting the ear to receive certain vocal or other sounds. . . . . . This power of adjustment results from the mode of attachment, position, and *voluntary* action of the stapedius muscle, affording to my conception a most close analogy to the ciliary muscle of the eye, especially when, as in predaceous birds, that muscle is inserted into osseous plates in the sclerotic, as strong and massive as are the crura of the stapes." I am always much assisted in interpreting the meaning of any given functional auditory disturbance, by bearing in mind the close analogy which appears to me to exist between the respective offices of the component parts of the ear and the eye. I have never seen any technical description of these in their relations, nor have I as yet fully promulgated my own views on this most interesting and instructive question. To enlarge upon them here would not be practicable, but until the resemblances at least in function, between the two organs of vision and hearing are recognized and worked out, no parallelism between their diseases can be well instituted; and it further seems to me that we must become convinced of the pathology of the two organs being also analogous, before we can hope to fill up the "great gap in our knowledge," truly so named by Von Tröltsch. If the well-known natural action of any part of the human organism, whether a muscle, a bone, or a membrane, be interfered with or annihilated, we are thereby at once directed to the exact locality of the morbid change. Reasoning from this fact, and recognizing the functional analogies between the eye and ear, we are guided to define more accurately the seat of disease in the latter organ, and thus to supply the deficiency implied by the authorities before quoted, when they state that 'a given impairment of function," or that "symptoms during life" do not enable us to distinguish which end of the ossicular chain is the seat of a serious affection, or even to identify so formidable an ailment as anchylosis of the stapes.

In my second lecture I attempted to give you a physiological explanation of "listening;" and why some persons, whose audition is defective, hear better amidst noises (pages 27 to 35). It was there described how the two little muscles, the *tensor tympani* and the *stapedius*, by their action, both separate and combined, regulated the tension of the membranes that are made to vibrate, and in conjunction with the ossicles, act as the analogues of the iris and ciliary body of the eye. The tensor tympani regulates the admission of sounds into the drum, by its influence on the membrana tympani, just as the circular muscular fibres of the iris (the sphincter pupillæ), by their contraction, diminish the size of the pupil, and prevent too powerful rays of light from entering the interior of the eye, and affecting injuriously the optic nerve. Following out this comparison, the *stapedius* muscle, by drawing the stapes outwards from the fenestra ovalis, relaxes the fluid of the labyrinth and the membrane of the fenestra rotunda, enabling them to become impressed by the most delicate vibrations and the faintest sounds. We now know, chiefly through Mr. Bowman's researches, that it is the *ciliary* muscle which is mainly instrumental in altering the form of the crystalline lens so as to arrange and adjust the eye for viewing near objects. This functional act in the eye (called variously *adjustment*, *adaptation*, or *accommodation*) finds its parallel in the ear in the contraction of the stapedius muscle; the perilymph and the membrana fenestræ rotundæ furnishing, in my opinion, the analogue of the lens. The analogy extends even to the nervous supply of the muscles, the stapedius and ciliary muscle being each excited to action by voluntary motor nerve-filaments, requiring sometimes — *e. g.*, in listening to an indistinct, monotonous speaker, or poring over minute objects of vision — so long sustained an effort of the Will as to fatigue sensibly whichever organ is exercised. The other two muscles under consideration in the ear and eye respectively — the tensor tympani and the circular fibres of the iris — are excited to contract in a different way — viz., by reflex action through involuntary or sympathetic nerves, which is called forth principally by the stimulus of sound or light in a manner well known to you.

It follows, that in attempting to determine with exactitude, in any given case, which part of the tympanic cavity is especially implicated in the catarrhal process, we have to study most carefully, in addition to the diagnostic evidences already explained, what particular auditory *function* is lost or interfered with. For instance, suppose a patient presents himself who is not very deaf, since he may hear tolerably well what is said to him while you speak slowly and distinctly, but he complains of being unable to follow general conversation, and requires to make a sustained effort of attention even when he is addressed by a single voice. We know that this one has lost the power of rapidly *adjusting* or *adapting* his ear to the tone of the speaker's voice. Or again, as I sometimes find to occur with persons who have cultivated a high power of musical discrimination, he has become unable to detect, as heretofore, any slight departures from correct tune and time in an orchestral performance.* If, in such a case, on examining the membrana tympani by inspection, &c., we find it tolerably healthy, and properly moveable under inflation through the Eustachian tube, and that there is

---

* The pitch of sound depends entirely upon the number of vibrations of a sounding body, the more rapid the vibrations the higher the pitch, and *vice versâ*. "Reciprocation of sounds" is associated with various degrees of tension of the tympanic membrane.

no evidence or history of nerve lesion, we may infer that the stapes has become restricted in its movements, or abnormally fixed in the fenestra ovalis, whether by rigid, dense, or thickened mucous membrane, by stiffened exudations or some of the many varied forms of adhesion which result from chronic catarrhal inflammation. This condition (which has been described on pages 28 and 142) I have termed "*impaction of the stapes*," and its effects are perfectly explicable by the fact that vigorous and sustained action of the stapedius muscle is requisite to move this, the last bone of the ossicular chain, and to keep it in the continued motion that regulates the pressure on the labyrinth fluid, and the tension of the round fenestral membrane through which all the finer and more delicate sound-vibrations must pass to the auditory nerve. We know that it is the contraction of the stapedius muscle, and not of the tensor tympani, which is needed thus to influence the labyrinth structures (the aural lens), because the prolonged effort of attention thus exercised causes such fatigue as to be sometimes insupportable; and furthermore, because unless this act of the will is sustained the sounds of a voice, or the time and tune of musical sounds cease to be perceived.

Thus much by way of proof that Politzer's "diminished mobility of the stapes" *can* be distinguished from "an adhesion of the head of the malleus to the upper wall of the tympanum." Were further illustrations wanted to prove conversely, that the latter disease is distinguishable from the former, and also from others, I might adduce the fact that it is next to impossible for such a condition to exist without impairing the free motion of the membrana tympani when subjected to inflation from within or to suction from without; nor would its appearance, colour, lustre, &c., remain unaltered. Certain sounds or tones would be heard, and others excluded: such vibrations only would be perceived as were able to be "reciprocated" by the degree of tension in which the membrana tympani was held by the immobile malleus. The contractions of the tensor tympani would be nullified, because the muscle would be unable to move the bone; but no fatigue would be experienced in the repeated attempt to do so, owing to the different mode of innervation of this muscle as contrasted with the stapedius. Many other physiological and pathological facts could be brought forward to show that affections of particular parts of the typanum are as easily and as certainly distinguishable as are diseases of the iris from those of the lens in the eye. I repeat, that analogical anatomy, physiology, and pathology will be of essential aid to us in the diagnosis of aural catarrh, and I suggest it to you as one of the most attractive as well as profitable studies to which you can possibly devote yourselves.

Another interesting symptom, which has been said by Mr. Toynbee,* and repeated by Mr. Hinton,† to to be characteristic of the later stages of "anchylosis of the stapes, or rigidity of the mucous membrane of

* Toynbee's "Disease of the Ear," page 281.
† Hinton "On Disease of the Ear," in "Holmes's System of Surgery," edition 1870, **page 310.**

the tympanum," is, *the hearing better in a noise*, as while riding in a carriage over a hard road, or during the beating of a drum. The first-named writer attributes the improvement in hearing manifested under these circumstances to "considerable vibration communicated to the body, which shakes the chain of bones and imparts to them a kind of vibratory movement, permitting the muscles while it lasts so to act on these bones as to restore more or less of their proper functions in adjusting the pressure on the labyrinth." The latter author does not explain the cause of the phenomenon, but admits it to be "common to this with some other morbid conditions." I myself cannot conceive it possible for bodily shakings and sonorous vibrations like the above to affect so decidedly a rigid ossicular chain and mucous membrane without at the same time injuriously influencing the more delicately organized and more vibratile parts of the labyrinth, such as the perilymph and round fenestral membrane, so as to produce utter confusion of all intelligible sounds. I am obliged to consider the foregoing interpretation of the very common symptom "hearing better in noises" as entirely erroneous; and (to my mind at least), it simply illustrates the necessity of bringing physiological reasoning to bear on pathological facts. You must not from this symptom diagnose the existence of any such severe catarrhal affection as "immobility of the stapes and rigidity of the tympanic lining mucous membrane." But you may assume that morbid structural changes have occurred at the other and outer end of the ossicular chain where the other muscle of the drum (the tensor tympani), which is excited to action by the stimulus of these noises, can be influenced. It is the membrana tympani which is the structure principally affected when the symptom in question is prominent. In the second lecture (pages 28-35) I have attempted at some length to describe the conditions and states of disease producing the phenomenon of "hearing better in a noise."

We have now, by accurate observation and analysis of symptoms, and the careful use of instrumental aids, ascertained (it may be supposed) the seat and description of chronic catarrhal disease in the middle ear, or, to speak technically, we have formed our *diagnosis* or *recognition* of the ailment. The next step is to determine, if possible, whether the resulting deafness is completely curable, whether it admits of relief by a slow or by a comparatively speedy process, whether we can hope by treatment to arrest the advance of disease, and so retain such hearing power as may still exist, or finally whether the probable ultimate event will be complete and hopeless loss of hearing. These are all points of extreme importance to the patient, and upon which you may often be eagerly questioned by him. This estimate and foreknowledge of the career and termination of a disease is, as you know, its *prognosis*, and varies, of course, with the almost infinite variety of cases — yet, by correct diagnosis, founded on a thorough acquaintance with the nature of the malady itself, you will be enabled in the great majority of instances, to arrive at it with accuracy, even in such an obscure and uncertain class of affections as we are now considering.

TREATMENT *of chronic aural catarrh.*

I now pass on to speak of treatment, which will consist chiefly in altering the morbid condition of the membrana tympani, the mucous membrane of the tympanic cavity, Eustachian tube, pharynx, nose, &c., and in restoring as far as possible, the right position of the ossicula auditûs : likewise, in attending to the general health of the patient. To take these particulars in regular sequence as heretofore :—

The *membrana tympani.* We shall seldom find its morbid appearances to be due to chronic inflammation of its structure alone, for this state will be either associated with or have resulted from a diseased condition of the cavitas tympani. The treatment will be the same in both instances. When there is thickening and general hypertrophy of the fibrous laminæ (a state frequently connected with the gouty or rheumatic diathesis) we should, in addition to local applications, prescribe such constitutional remedies as are efficacious in gouty or rheumatic ailments. With a hope of clearing off the opaque spots which stud the surface of the membrane, it is recommended to apply, with a camel's hair brush, or on a piece of cotton wool wrapped round a probe, a solution of the nitrate of silver (grs. v—xx to ʒj. of water).

Some darkish scales of dermis peel off, and occasionally we find the membrane thinner, and a little more transparent after repeated applications. The deafness, however, not being dependent upon the condition of the membrana tympani alone, is seldom thus benefited, but annoying tinnitus may sometimes be relieved. A few drops of sulphuric or acetic ether, or of spiritus camphoræ combined with liq. morphiæ, ᴣj. of each, to ℥j. of glycerine, will occasionally answer the same purpose.

In relaxation, thinning, or atrophy, when the membrane may look like crumpled paper, and depressed spots, resulting from some loss of its substance, are observed, astringents, such as zinci sulphat: (grs. v—xx ad ℥j.) liq., plumbi. diac: ᴣj. ad ℥j., &c., poured into the ear, and allowed to remain for a few minutes, do good. I also recommend, for the purpose of relieving the troublesome "noises," an application composed of equal quantities of atropiæ sulph: and liq. morphiæ hydrochl: But the best remedy where the deafness is considerable, and there is reason to suppose that the ossicula are disarranged or dislocated by the previous catarrhal processes within, is Yearsley's "artificial tympanum."* This, when carefully adjusted, will in some cases give instant and marvellous relief to the deafness. It acts by supporting the ossicles, and maintaining them in a position to convey vibrations across the tympanum, and give thereby the needful degree of tension to the membrana fenestræ rotundæ.

I could adduce from my note-book several instances where hearing was quickly restored, and afterwards preserved by the insertion of this small

---

* This is not to be confounded with Toynbee's appliance similarly named, which consists of a piece of gutta-percha tissue, harsh and often irritating, mounted on a stalk.

plug of cotton against a thin atrophied membrane which had been driven outwards by the injection of fluid into the tympanic cavity. In these cases the surgeon consulted had not sufficiently recognized the attenuated condition of the membrane, but had employed his usual treatment for tympanic disease, where it was assuredly contra-indicated; and almost total deafness had been the unfortunate consequence.

In the greater number of cases where "collapse" of the membrane has occurred, and it has accordingly fallen inwards towards the opposite wall of the tympanum, the Eustachian tube has been obstructed or altogether occluded. In such instances, no external application will be of any avail; but perseverance in treating (as will presently be described) the impervious Eustachian tube and contracted tympanic cavity which have caused the "collapse," will often be followed by recovery of hearing.

The management of otorrhœa and other discharges from the surface of the membrane and from the auditory canal, which occasionally attend chronic catarrhal inflammation, will be detailed under the head of purulent aural catarrh, or otitis.

Before proceeding to lay down rules for the local treatment which appears to me indispensable for the relief of deafness depending upon obstruction of the Eustachian tube and cavity of the tympanum, I will read a translation from Politzer's admirable treatise,* giving a brief abstract of "The Treatment

---

* On "The Membrana Tympani in Health and Disease." American edition. By Drs. Mathewson and Newton. 1869.

of Chronic Catarrh of the Middle Ear and Eustachian Tube, without Perforation of the Membrana Tympani." You will thus be acquainted with the mode considered as the most effectual by the aural surgeons of Germany.

"After first inspecting the meatus and membrane we determine the hearing distance, both for the watch and the voice, and then proceed to the examination of the Eustachian tube, forcing air through it into the cavity of the tympanum in the way before specified (Politzer's new method), or by the catheter, and determining by means of the otoscope whether a current of air enters the cavity. When the hearing distance is hereupon noticeably increased,—an inch or a foot for the watch, and several feet or fathoms for the voice—and we may infer therefrom the existence of swelling and excessive secretion from the mucous membrane of the tympanum and Eustachian tube, an injection of a solution of zinc into the middle ear is indicated, besides the employment of the 'new method.'"

"From four to eight grains of the sulphate to the ounce of water may be used. The catheter is introduced into the Eustachian tube, and held with the left hand. Some of the astringent solution is put into it by means of a little glass tube, and is blown into the tympanum by compressing with the right hand an elastic bag attached to the catheter. These injections should generally be repeated every third day, and continued for from three to five weeks. If, as in many chronic cases, no complete recovery, but only more or less marked improvement, takes place; the injections, after some months' interval, should be renewed for two or three weeks, to prevent, if possible, the further increase of deafness."

"If, on the other hand, there is little or no increase in the hearing distance, after repeated employment of the air-douche, we may infer that the deafness is caused by the sequelæ of the catarrhal affection, viz., thickening of the mucous membrane and of the covering of the ossicula, with rigidity and diminished mobility of them, we can only expect improvement from the use of moderately stimulating injections, together with the air-douche. The following solutions seem best suited for injection:—Ammoniæ hydrochlor: gr. xx.; potassii iodid: gr. x.; sodii chloridi: gr. v. to the ounce of distilled water. After the employment of these injections from two to four times a week, in the way before described, we shall in some cases obtain essential improvement in the hearing; in others, the improvement is only slight, but we shall occasionally overcome, at least, the tinnitus, vertigo, and confusion in the head, which are symptoms accompanying chronic thickening of the mucous membrane of the middle ear. The improvement obtained by these saline injections in the cases supposed, seldom continues, since the thickened tissue of the mucous membrane, which was somewhat softened by the air-douche and the stimulating injections, has, like cicatricial tissue, a tendency to retract, whereby the rigidity of the ossicles returns. It is necessary, therefore, to repeat the injections and air-douche from time to time (for instance, every three to six months), every other day for a fortnight or month. The introduction of elastic bougies into the Eustachian tube is sometimes attended with essential improvement. We must remark that it is of especial importance to remember that a constant or too protracted employment of these irritating injections is injurious, while a treatment interrupted by pauses of weeks or months proves most effective."

This line of treatment, adopted I believe, by one English aural surgeon and by most of those on the Continent, is in my opinion too violent in character, and too abruptly entered upon. My own experience is opposed to the sudden and early introduction of the catheter, either as being necessary to diagnosis, or as a therapeutic agent. I would advise your resorting to gentler and less disagreeable or painful modes of opening the Eustachian tube, and restoring air to the drum cavity; neither can I approve of the sudden and forcible injection of fluids. In a certain class of cases which have been brought to me, in which this mistaken treatment had been pursued, the hearing was very disastrously affected, and in one instance, not recoverable even to its former standard.

In *Obstruction of the Eustachian tube*, whether this condition arise from thickening, relaxation, or any other morbid condition of the faucial mucous membrane, or from disease proceeding outwards from within the tympanum, our first efforts must be directed to restore the perviousness of the canal. Place the otoscope in the patient's ear, and direct him to endeavor to blow air into the drum, by holding the nose, shutting the mouth, and making a forced expiration (the Valsalva method).* This manœuvre need not be many times resorted to, for if, after a few attempts, it prove ineffectual, as it sometimes is with young children, and even some grown persons, from their sheer inability to understand the process, recourse should be had to Politzer's beautifully simple and most valuable invention.† The modified contrivance which I have so recently introduced obviates the necessity for passing any tube up the nostrils, and this (I think) enhances the practical value of Politzer's discovery. Children do not cry, nor do "nervous" ladies become frightened at this form of inflation, for it is not only painless, but neither uncomfortable nor alarming in its application. Its results are in many cases at once successful and astonishing to the patient.

The sounds heard through the otoscope, and the alterations in appearance of the membrana tympani during and after inflation by either of the methods above described, and by catheterism, have been so fully detailed, and also frequently tested by yourselves in the out-patient's room, that you cannot help being familiar with them. They will generally indicate the condition of the tympanum, and whether or not there be fluid in its cavity. As Eustachian obstruction is of much more frequent occurrence

---

* For description of this method, see pages 66 and 74, and for cure by it, page 14.

† Explained and illustrated, pages 78–81.

‡ See Cases III. and IV., described at the end of this Lecture, where most striking effects were produced, after fourteen and seventeen years' duration of deafness.

in children than in adults, and in them is easily overcome, the employment of the catheter for removing it has now been rendered almost wholly unnecessary; in fact, I have not once had occasion to use that instrument upon a child under fifteen years of age, since Politzer's discovery became known to me.

If air cannot, by the gentle means specified, be passed into the tympanum, a careful diagnosis of the physical cause of the impediment should be made, and if it is ascertained to be due to a thickened, relaxed, or generally diseased condition of the mucous membrane of the throat, nose, pharynx, &c., which has continued itself into the faucial end of the tube (a frequent source of Eustachian obstruction, as you are aware), both local and constitutional remedies should be adopted before you resort to the catheter. Enlarged tonsils, nasal polypi, adenoid growths, &c., all of which have been previously specified as likely to co-exist with occlusion of the tube, and to act as predisposing, if not exciting causes of obstruction, must be treated at the commencement.

If the tonsils be very large, granular, hypertrophied, subject to small but recurring abscesses, and in consequence of their size, press the soft palate out of position, so as to restrict the proper action of the throat muscles which open the tube, I advise you to excise them without delay; I mean by this, that you should remove such portions as project beyond the palatine arches—not the whole gland, for that which remains after excision will contract and shrink up. Even if the enlarged tonsils have not yet injuriously affected the hearing in the indirect ways before specified, their removal is desirable, on the ground that they seriously interfere with respiration, deglutition, and speech; they likewise damage the general health, and retard the development of the chest.

"Local applications to the enlarged tonsils are often recommended, and too constantly adopted. Nitrate of silver rubbed over their surface, or points run into the substance of the glands; sometimes nitric acid carefully applied to portions of it; stimulating gargles, and a variety of useless and troublesome applications, have had their advocates for the arrest or removal of the masses, but it appears to the author a useless waste of time and material to attempt to procure absorption of such dense tissues. If constitutional treatment does not arrest, and local treatment be necessary, removal by the knife of a portion of the gland is the speediest and in our opinion, the only efficient remedy. The surgeon with a knife can remove all that is necessary in a few seconds, and the patient will be rid of the consequences in less than a week afterwards."*
Sometimes it is necessary to cut off an elongated and hypertrophied uvula.

Local remedies to the throat are now to be considered. We must never neglect to treat the mucous membrane of the fauces by special applications, for

---

* Mr. G. D. Pollock, in "Holmes's System of Surgery." Vol. IV., page 84.

it is possible that by such means we may succeed in reducing the thickening or granular condition, so as to allow the tube-muscles to resume their function; when, on the tube being opened, the deafness will suddenly disappear, with a "pop" or a "crack" as patients say. Such a desirable event must not, however, be generally expected. It must be looked upon as fortunate should the local remedies be successful in removing the granulations or other outgrowths, and in so far modifying the thickened or diseased condition of the mucous membrane surrounding the tube, that on a repetition of the "inflating" process, the obstruction shall yield. For getting rid of the adenoid vegetations, we may perhaps find it necessary to adopt operative measures, which will consist in crushing or scraping them off by appropriate instruments: or should they be very soft and small in structure, by *cauterisation* with the solid nitrate of silver. All kinds of granulations and condensed mucous structure are best treated with caustic, either in the solid form, or a strong solution, from gr. 20 to 60 to an ounce of water. Use Toynbee's or similar caustic holders, which will allow of the bent end being passed behind the arches of the palate, so that the caustic may touch the orifice of the Eustachian tube. It is never necessary to introduce this agent through the catheter, as is recommended by some high German authorities who appear determined to poke that instrument into the nostrils upon the slightest possible excuses.

In almost all abnormal states of the throat causing congestive swelling of the parts in the neighbourhood of the Eustachian tube, solutions of nitrate of silver are undoubtedly the most efficient local applications, and they are those best borne by the patient. It is not too trivial to mention here that bent whalebone with a piece of sponge attached, or a large camel's hair brush turned at an angle, are not the appliances best adapted for touching the faucial openings of the tubes. In using them, the fluid is wasted on the front surfaces of the soft palate, instead of reaching the pharyngeal space behind. I always use a strong silver spatula, curved at the thin flattened end, round which I wrap cotton wool thoroughly saturated with a solution of nitrate of silver (40 grains to the ounce). Perchloride of iron, diluted with glycerine, may sometimes be substituted for the caustic lotion, or powdered alum, blown through a bent tube into the space behind the tonsils, is a useful astringent.

I am in the habit of recommending an elastic bottle and tube for the purpose of washing the nostrils, fauces, &c., which can either be introduced through the nostrils, or when curved at the end, passed into the mouth so as to inject the lotion in the form of spray against any affected parts. Vitiated mucous secretion can by this means be daily washed away by the patient, and the lotion will be brought into contact with the secreting surfaces. The effect produced upon the nasal and faucial passages is much more satisfactory than by gargling. This appliance seems peculiarly adapted for the cases of children, who do not, in general, easily learn to gargle. By combining with the injections carbolic acid solutions, or Condy's fluid properly diluted, the

200   CHRONIC CATARRHAL INFLAMMATION;—

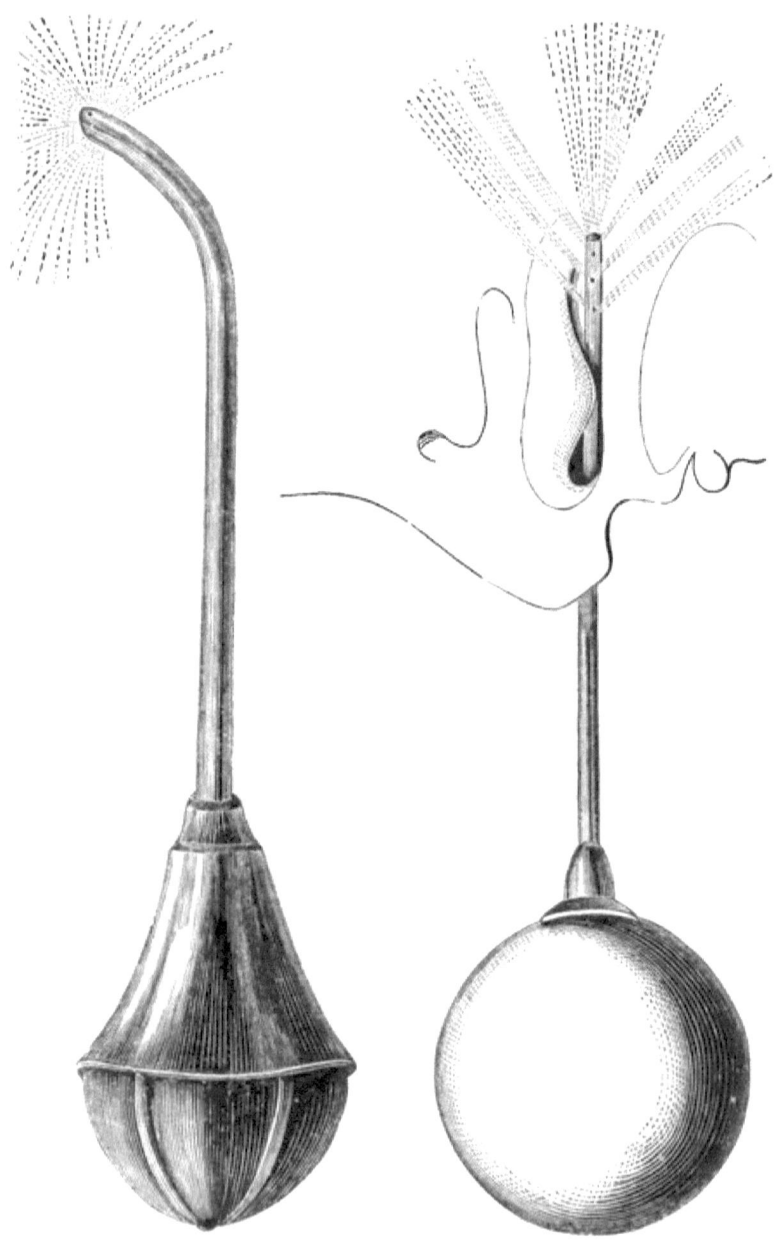

fœtid odour of the breath which so often attends these troublesome collections of vitiated stringy mucus will be lessened, if not wholly prevented. Astringent, alkaline, iodine, absorbent, sedative, or any other kinds of lotions may be employed through this instrument. (*Vide* engraving on the opposite page.)

If all the means just described have been tried in vain, and the Eustachian tube remains obstinately impervious to air, we are compelled to have recourse to the catheter. The kind of instrument to be preferred, the mode of introducing it, the precautions requisite in its manipulation, its utility and comparative value in the diagnosis as well as treatment of catarrhal affections, the effects of the air douche and fluids administered through it, and the sounds heard in certain diseases at the time through the otoscope —have all been so fully detailed at pages 83—98, that I am not obliged to break off from the subject of this lecture for the purpose of describing them. Two deaths occurred about thirty years since in Dr. Turnbull's practice, which made a great sensation at the time in London, and brought Eustachian catheterism into very undeserved ill-repute. Mr. Hinton is of opinion that one of them was due to "extravasated air obstructing the pharynx; but this cannot be made certain, and the evidence proves that the air was injected in a very rash and violent way, and" (the operation) "was entrusted to hands quite unskilled."

I have always maintained that a pervious state of the Eustachian tubes is absolutely essential to perfect audition, even when the rest of the middle ear is apparently free from catarrhal disease; but the following case would seem to furnish undeniable proof that their patency must be insured, notwithstanding that air can be freely admitted into the cavity of the drum through a lost or perforate membrane. In this instance I think that the absence of any sufficient means of escape for the tenacious clogging secretions was the chief cause of the severe deafness; for I have often noticed when adjusting the cotton-wool "artificial tympanum" in cases of perforation attended with otorrhœa, that I fail to produce the immense improvement in hearing, which is effected subsequently when the patient inflates the tympanum and blows the secretion out. I imagine that thick films of muco-purulent secretions adhere to the fenestræ, especially to the round opening, and of necessity impede vibration.

The particulars of this case will be only briefly related, for it is not one of chronic aural catarrh, but of the purulent form, which will be the subject of our remaining lecture. The patient is a young gentleman twelve years of age, who became deaf nine years ago from an attack of scarlet fever. For the last two years he has been under the care and treatment of a well-known aural surgeon, who perscribed constitutional remedies and some lotions for the ears, but had never inquired into or tested the perviousness of the Eustachian tubes. No benefit had followed this treatment. When I first saw this lad (who was, I may mention, on account of his severe infirmity, inferior in intelligence to his younger brother, who came with him to my house) he could only hear my watch in close contact with his right ear, and but one inch from the left. The otorrhœa was copious, purulent, and offensive. After the discharges had been carefully syringed away there was but little perceptible increase of hearing power on either side, and an inspection revealed almost entire loss of the right membrana tympani, and a large perforation in the left. The mucous membrane of the cavitas tympani was granular, and there were small fungoid growths in the left meatus. While syringing with warm water, none passed into the throat, as might have been expected to occur from the extensive loss of membrane and the consequently open state of the tympanum. The patient not being able himself to overcome the obstruction thus plainly existing in the Eustachian tubes, by the Valsalva experiment of inflation, I employed my modified Politzer bag, and succeeded easily in blowing air and some of the purulent secretion through the tympanum into the external meatus. The emission of both air and fluid with a whistling and hissing noise from the middle ear, of course still further assured me of the existence of the perforations. The boy's hearing distance immediately rose to 2 inches on the right side and 6 feet on the left. 2nd visit: The watch was heard 3 feet on the right side and 10 feet on the left. 3rd visit: 12 feet right, 14 feet left; and so on, improving rapidly each time, until at the sixth visit it had increased to 25 feet on each side, or the whole length of my consulting room, at which distance my watch continues to be heard whenever he visits me.

It is not too much to say that the comparative restoration of hearing in this case depended *entirely* upon the removal of obstruction in the Eustachian tubes. Of course, the hearing is not even now perfectly restored (although, to hear any watch at 25 feet distance must be considered hearing *well*), because of the extensive loss of structure in both ears; but I shall be able to maintain it at its present standard by the use of the artificial tympanum.

When the Eustachian tube is rendered impervious by dense stricture, by fibrinous effusion uniting its walls, or by very inspissated mucus, the patient should not, in my opinion, be made to endure exploring by bougies passed through the catheter, unless there is very strong evidence to prove that the deafness certainly arises from the occlusion of the tube, and not from either tympanic or nerve disease. Should every means of inflation fail, it may be necessary to dilate the tube mechanically; but such an operation is very seldom really required. Mischief has of late years often resulted from the practice of attempts to pass stilettes, whalebone, catgut and laminaria bougies into the Eustachian tube. I have, in several instances, been perfectly successful in attaining the desired object simply by perseverance in using the local applications before indicated, and by forcing air *into* the tube by catheterism. A gentleman is at present under my treatment for chronic catarrhal affection of the tympanum, into whose Eustachian tube one German aurist repeatedly attempted to push a catgut bougie without beneficial effect. He can now, however, inflate that ear, in which the disease is of ten years' standing, even more rapidly than the other, which does not appear seriously obstructed. In this case an increase in the hearing power has followed upon the cavity of the tympanum being able to be brought under the influence of the air douche, &c.

Where the deafness is occasioned by a *chronic catarrh of the tympanic cavity*, and the Eustachian tubes are quite pervious, the first point is to determine as precisely as possible the exact nature of the lesions. One of the most frequent causes of deafness, especially in children and young persons, is a *collection of mucus or other extraneous matter* within the cavity of the drum. Means must be adopted for causing its absorption, or removal by the air douche, and perhaps by fluid injections. Good management of the general health, and appropriate local treatment, will in most instances, cause these accumulations to disappear. With the object of correcting the altered condition of the tympanic cavity, and of securing quite a free passage of air to it, the air-douche is most useful. Politzer's method, with the modified bag and apparatus which have been so often referred to, may be sufficient for the purpose. A mechanical pressure will in this way be brought to bear upon those parts of the tympanum which are elastic,—the membrana tympani and the fenestræ. This distension will have a tendency to relax these parts, and at the same time may counteract incipient rigidity of them and of the ossicles, and will loosen any recently-formed adhesions between these and the walls of the tympanum. When the Eustachian tube has once been rendered, by these means, completely patent, it is well to advise the patient to maintain it so by practising the Valsalva method of inflation at least once a day. If children, in whom there is often merely a tubal obstruction with a little hypersecretion of mucus in the tympanum, can be taught to keep the passage free to the middle ear by this simple action of holding the nose and blowing air into the drum, nothing more is necessary for removing their recurring deafness at every catarrh; and in adults the improvement or recovery of hearing is generally maintained by the process. If they cannot manage it themselves, direct their attendance occasionally for treatment by the inflating bag.

Before quitting this subject of accumulations, let me seriously warn you, in dealing with such cases, not rashly to fly at the knife and lay open the tympanum, brilliant though such an operation may appear, and instantaneous its beneficial effect on the hearing. Sir Astley Cooper first introduced it in this country, but Sir William Wilde correctly says of it—"There has not been, perhaps, in the whole history of medicine during the present century, a discovery to which so much praise was at the time awarded, that subsequent investigation and experience have, to say the least of it, so much disparaged." I need scarcely remind you that the great English surgeon above named, latterly discarded "puncturing" or as it is now called, "incising" the membrana tympani, even in those cases, to which he at one time thought it applicable. But the operation of " incision " or " paracentesis " has lately been revived, principally by Dr. Hermann Schwartze of Halle,* and rather extensively (in some instances I think very rashly) adopted in London. I have myself witnessed cases in which its evil results have been conspicuous, and permanent deafness with difficulty averted; and I am strongly of opinion that the circumstances in which it may fitly be resorted to, are of very rare occurrence. Observe, I must be understood as applying these remarks only to the treatment of *chronic catarrhal* effects, and not to cases of purulent catarrh, or *otitis*, where a collection of pus ought, under all circumstances, to be evacuated from the tympanum as soon as its presence there is ascertained. In cases of the former class, the improvement in methods of diagnosis, and in facilities for treatment by the air-douche, and by catheterism with suitable fluid injections, will it is to be hoped save you from ever needlessly performing a hazardous operation, and thus bringing discredit on the practice of aural surgery. Case No. 2, at the end of this lecture, will illustrate the warning I intend to convey.

When the too great secretion within the tympanum is not removed by the air douche alone, administered with the bag,† warm vapours suitably medicated should be injected into the cavity. In most cases which are of some months or years' duration, the thickened, swollen, and over-secreting lining mem-

* See Archiv für Ohrenheilkunde, " die Paracentese des Trommelfells," 11, IV., s. 264 (1868).
† It must be remarked, that so long as a moist gurgling sound is heard through the otoscope when air is forced into the tympanum, excessive secretion may be inferred.

brane of the drum requires to be decidedly acted upon; and I find moist vapour (with which I sometimes mix acetic ether) to be the most efficient agent in softening the rigid hypertrophied tissues; and also in assisting the absorption of the effused fluid. Other vapours, such as those from iodine, carbonate or hydrochlorate of ammonia; chloric ether, henbane, conium and other sedatives, may be cautiously used, but the effects of many of these can only be ascertained after several administrations. If one kind does not succeed, another may. I have employed these and other similar vapours with varying success in the treatment of chronic catarrh, but my experience nearly agrees with that of Dr. Von Tröltsch, that with the exception of iodine and acetic ether, the vapour of hot water—*i.e.*, simply steaming the cavity of the tympanum, is the best. I would myself leave out iodine, for I have observed that its tendency is to irritate too severely all the mucous membranes with which it comes in contact; and in one instance I think it induced an attack of asthma. The patient was however liable to that complaint. The injection of any of these vapours, except the irritating ones, causes in general no pain; but when the cavity has become diminished in size, from the congestive swelling, deposits, adhesions, &c., the mechanical distension of some parts and the breaking up of adhesive bands will cause a biting or aching pain, not however, of long duration. The treatment should be steadily continued—for several days—and many cases which have been considered beyond remedy will become immensely relieved or even cured. You have opportunities of noticing how perseveringly some patients who have been deaf for ten, twenty, or more years, return week after week to this hospital for treatment, and to have catheterism performed with signal and increasing benefit.

Where accumulated mucus has become dried and very tenacious, the operation just now referred to, of incision or paracentesis of the membrana tympani, has again been advocated, to be followed by blowing the retained secretion into the meatus with the Politzer bag, or washing it out by the injection of various alkaline solutions through the catheter. It is the practice of Mr. Hinton "to syringe a strong solution of soda through the tympanum by the meatus, on each of the four days subsequent to the incision, if no irritation arises."

Now, Dr. Schwartze, who as you recollect, first brought the practice of incision into its renewed vogue, advises it "in certain cases of acute inflammation of the membrana tympani, where in a very short time a great swelling occurs, chiefly and greatest in the posterior and upper quadrant, and when in spite of other remedies there is a very obstinate and severe pain. It quiets the pain, and thus shortens the course of the affection." In this description you cannot help recognizing the earlier stage of otitis, or, as I term it, *purulent catarrh*. But Schwartze himself decidedly disapproves of incision where there is simply tenacious mucus in the cavity of the tympanum, because he fears that the small collections are never evacuated through the opening, and that "very severe reaction often occurs in such cases after the paracentesis, active inflammation which terminates in suppuration, and often with considerable diminution in the acuteness of hearing." Von Tröltsch, in speaking of incision being performed in cases of *chronic catarrh*, says, "Among the many reports of favourable results from this operation, none of them could be said to give any sufficient evidence of its real value; and I must, therefore, fully coincide with Schwartze, when he says that 'up to the present time it is only in very rare cases that a permanent success has been seen by trustworthy observers.*'"

Where the mucus has become dried and hardened in the tympanum, impeding the passage of sound, and setting up adhesive processes which may at a later period act most disastrously upon the hearing, no time must be lost in endeavouring to modify so formidable a condition. I believe it is frequently these cases which are set down as "nervous deafness," "thickening of the drum," "anchylosis of the stapes," "rigidity of mucous membrane," "bands of adhesion," &c., &c. Improvement is generally effected with ease, by injecting through the catheter weak alkaline solutions, which will first soften, and then dissolve the hardened masses. Afterwards, when sufficiently thinned, and rendered less irritating

---

* You have probably all heard of, and some may even have seen paracentesis abdominis performed when the real ailment, or rather condition, was one *more natural and less alarming* than either ascites or ovarian dropsy. Such a mistaken operation was called by Sir Astley Cooper, " dry tapping !" If statements may be relied upon, paracentesis membranæ tympani has, on some occasions, likewise proved " dry."

and clogging, the altered secretion may be treated in the same manner as above recommended for simple mucous accumulation. As solvents, I usually employ injections of the bicarbonates of potash or of soda, or iodide of potassium, and afterwards as an astringent, sulphate of zinc (gr. ij.—v. ad. ℥j.) It is well to suspend this treatment for a few days, and employ only the air-douche with steaming. The lotions may either be placed in the nostrils by a syringe, and then blown into the tympanum with the Politzer bag, or if not objected to, the catheter may be made use of for the purpose. For very chronic cases, and where the patient is unable to give regular attendance, Dr. Gruber of Vienna has pointed out a simple method by which the patient can himself introduce fluids into the tympanum. " Inclining his head well to the affected side, he puts into the nostril on the same side a glass or elastic syringe, containing about ʒij. of the required solution. This he syringes firmly into the pharynx, where it lies in contact with the orifice of the tube. He then quietly, and without swallowing, inflates the tympanum."

Counter-irritation, by painting Lin: Iodi, P. B. on the mastoid process behind the ear, is very useful in treating chronic catarrh of the tympanum, especially when there is reason to believe that the mastoid cells are partially clogged.

The *constitutional treatment* of chronic aural catarrh can be considered only as auxiliary to the local, yet much stress should be laid upon the necessity of exercise, bracing air, warm clothing, &c., and upon the taking such medicines as impart tone to the system. In children particularly, who suffer from a generally disordered state of the throat, languid circulation, and consequent deficiency in nervous energy, great attention should also be paid to the skin, and to proper diet.

In cold weather warm baths are advisable, taken in the house, not out of it: in summer, cold—especially sea—baths, and vigorous friction with rough towels afterwards, so as to produce a ruddy glow. Plunging the head in cold water, as boys so frequently do after exercise, &c., is objectionable. . I have known severe catarrh to be induced by this practice, and bad cases of thickening to result from it, of course causing deafness. Youths should not be overworked mentally, and early hours for retiring to bed ought to be insisted upon. Sleeping-rooms should be well ventilated, and all general hygienic rules strictly observed. Cod-liver oil, various preparations of iron sometimes combined with alteratives such as hydrargyri bichlo:, mineral acids, vegetable bitters, iodide and bromide of potassium, will all be found useful by turns.

CASES.

CASE 1.—*Catarrh of middle ear (left side). Serous or thin mucous accumulation in the cavity of the tympanum, seen on inspection and felt moving about by the patient. Eustachian tube obstructed. Considerable deafness. Cured by Politzer's method of inflation.*

Master A. C., aged sixteen, was brought to me from Dorsetshire this summer. Deafness came on from cold taken after measles. His general health was not more than tolerable; he looked pale and had been rather overworked in studying for an approaching competition. On examination the throat was flabby, tumid and granular; tonsils enlarged; breathing more through the mouth than nose, snored when asleep. The hearing varied, sounds generally appearing to him muffled and deadened, or faint as if coming from a long distance. His own voice had an increased resonance. There was a sense of fulness in the left ear, and when he inclined his head, a feeling as if something was moving about within it. The tuning fork placed on the vertex was heard on the left (the deaf) side. Inspection showed the membrana tympani to be sunken inwards; thin and tolerably transparent, but darker in colour at the lower half. The processus brevis of the malleus looked much foreshortened, from its handle projecing into the tympanic cavity. This bone was extremely well defined. When the head was placed horizontally, the darker shaded portion of the membrane changed its position, appearing at the posterior half (then, of course, the most dependent part). From this symptom and the subjective one of the "something moving about," I inferred that there was an accumulation of serum or mucus in the drum cavity. The left Eustachian tube was impervious, and the attempt to inflate by the Valsalva method caused so much giddiness, that it was not repeated. On using the Politzer bag, air entered with a moist flapping and crackling sound very audible through the otoscope. When the membrane was again inspected, I observed it had almost returned to a proper position and curvature, and its colour had become more generally bright. A striking increase had taken place in the hearing distance, from two feet to fifteen for the watch, and the fulness, heaviness, and tinnitus immediately ceased. Nitrate of silver solution was applied to the fauces, and lin. Iodi. behind the ear; the latter with the object of promoting absorption of the fluid which the inflating process had doubtless dispersed into the mastoid cells. On the following day the patient was able to press air through the Eustachian tube by the Valsalva method, without experiencing giddiness. He had had no return of fulness in the head or singing in the ear, and the hearing distance was fifteen feet. I again used the Politzer bag, and the sounds indicating moisture were less audible than on the previous day; but as has been remarked the noise of the bag itself is apt to mask minor sounds.

Hearing distance afterwards twenty-five feet — or normal.

Catheterism in this case would have been wholly unnecessary. The local treatment, combined with tonic medicines, was recommended to be carried on for a few days longer. The patient has continued to hear well.

CASE 2.—*Catarrhal inflammation (subacute) of left tympanum; retained mucous secretion made to extrude through an aperture in the membrana tympani by means of the air douche, administered with the Politzer bag. Obstructed Eustachian tube on both sides. Recovery of hearing. Chronic catarrh on the right side, with extensive calcareous deposits in the tympanum and deafness of forty years' duration. Incision contemplated, catheterism with injections adopted. Hearing almost restored.*

Mrs. A., the wife of a medical gentleman, consulted me in this year (1870). I was furnished with a long history of the various modes of treatment adopted. The patient and her family were gouty, and a brother had been very deaf for the last twenty years.* This lady's infirmity was therefore supposed

---

* I have since been consulted by this gentleman; I found him to be suffering from obstructed Eustachian tubes and chronic catarrh of the tympanum. He could only hear the watch a few inches; but after being treated in the ordinary way (with the air douche, &c.) he has nearly recovered his hearing.

to be hereditary, and "nervous." Her health was evidently undermined by anxiety, and the fear lest the rapidly-increasing deafness on the left side should become as bad as that on the right, which had always been useless to her. She never remembered to have heard voices with the right ear. The left became deaf two or three years since; and is considered to have been worse after and during "colds." She had formerly had otorrhœa in the right ear, and about three months ago the same ailment in the left; tonics and various other medicines had been taken; medical friends' suggestions made trial of, but all to no purpose; the infirmity increased notwithstanding all constitutional treatment. The mucous membrane of the throat was congested; occasional darting pain was felt in the left ear, never any in the right, but continual singing noises there. She was unable to inflate the tympanum on either side, and the attempt produced giddiness; she required to be spoken to in a loud voice; and the powerful watch-tick (usually heard at thirty feet off) could only be distinguished at one inch on the right side and ten on the left. The right membrana tympani was studded with greyish-white deposits of crescent shape, which I considered to be the product of former otorrhœas or mucous collections which had passed away without perforating the membrane and had degenerated into chalky deposits (calcareous metamorphosis). After having removed some discharges of desquamated sodden epidermis from the left meatus, the membrana tympani on that side was observed to be rather dull, flattened, and of a pinkish hue in some parts; I thought I could distinguish a small aperture in the lower posterior segment, and the malleus was visible. The Eustachian tube being impervious, Politzer's method was employed. Immediately air entered the tympanum, the unmistakable hissing sound assured me of the existence of a perforation on the left side, and on inspection, a long string of tenacious mucus was seen lying in the meatus, which I removed. The hearing distance instantly rose to six feet.

During the remainder of this lady's stay in town, I adopted no further measures for treating the right ear. The Eustachian tube on that side remained impervious notwithstanding the use of the Politzer bag; and as I had no hope of relieving such long standing deafness and disease without evacuating the masses of chalky deposits so plainly seen lying in the tympanic cavity by incision of the membrane, I postponed doing anything more until her husband could attend. The aperture in the left membrane healed the following day; and the hearing power became reduced to three feet. Inflation for three successive days, however, soon improved the hearing, and she left town with her deafness greatly relieved. By attention afterwards to the state of the throat, and some astringent lotions applied to the ear, she reported the hearing as restored.

At the end of two months, Mrs. A. came again — this time with her husband — for further treatment. On the left side the watch was heard at sixteen feet, and the voice in a whisper. The right ear was as deaf as ever, and there was no alteration in the aspect of the membrana tympani. Before, however, proceeding to carry into effect the predetermined-on operation of "incising" the membrane, it was necessary to be assured that the Eustachian tube could be made pervious; accordingly, I passed the catheter. Finding air to enter the drum with a moist, rattling sound, I injected a little warm potash water rather strongly into the tympanum. Immediately the hearing improved; the watch could be heard at two feet, and the tinnitus lessened. On inspecting the membrane, the deposits within had partially cleared off, the lower half only being now covered with them. Of course, all idea of performing the contemplated operation was abandoned. The next day I found that the same degree of hearing had been retained. Catheterism, with injection was repeated; and more of the accumulated secretion was thus softened and washed away; hearing four feet. After the fourth day's treatment, eight feet of hearing-distance had been attained, and the appearance of the membrane had greatly changed. It was brighter, of nearly normal curvature, and there remained only one patch of the thickened secretion at the bottom of the tympanic cavity. This lady, I am informed, does not now consider herself to be "deaf at all."

CASE 3.—*Obstruction of both Eustachian tubes, from thickening of the faucial mucous membrane; slight accumulation of mucus in the tympanum; great deafness of fourteen years' duration. Cured by inflation with the Politzer bag at one visit.*

Mr. S., of Sheffield, aged eighteen. His mother

states, that he has been dull of hearing since four years old, when he "took a severe cold." The deafness always became worse now during a cold. He snores in his sleep, and keeps his mouth habitually open; has consequently a vacant, heavy expression of face. On examination, the mucous membrane of the fauces appeared to be thickened, the tonsils somewhat enlarged, and tenacious phlegm hanging about the throat; the nasal membrane was also thick. Inspection showed the right membrana tympani to be opaque and very concave, the left not quite so concave, but also opaque. The tuning fork was heard equally well on both sides; there were humming noises and a feeling of fulness in both ears. The watch was heard only four inches on the right side and two on the left. Eustachian tubes impervious. I used the Politzer bag, when air rushed into the tympanum with a gurgling noise, and into the left with a clapping sound, which seemed (to the patient) like the explosion of a gun. Both membranæ tympani were thus thrust outwards to a more natural position, and the "ground glass" appearance of the one was less observable than before. I repeated the inflation several times, until there was less of the moist gurgling sound in the right tympanum, where I had been led, by both objective and subjective signs, to conjecture that mucus was retained. The hearing distance for the watch rose quickly to twenty-five feet—the end of my room. The next day the same power was found to have continued; and on repeating the air douche, the sound of moisture was absent. I have lately been informed that the recovery of hearing has been perfect.

CASE. 4.—*Chronic catarrhal inflammation of the faucial orifices of the Eustachian tubes, causing complete obstruction and severe deafness from infancy. " Adenoid growths" or granulations in the naso-pharyngeal space. Cure.*

Mr. C. W. B., aged eighteen, articled to an architect, consulted me quite lately. His general health was not good; he looked pale, and was not strong. He generally breathed through his mouth, which was constantly open. Certain consonants were not pronounced; he would say "lo" and 'do" for "no," &c. His speech had what is called the "dead" pronunciation, and the voice was wanting in resonance. With the exception of these symptoms, so indicative of catarrhal affection of the naso-pharyngeal cavity, the case is very similar to the one just detailed, and it is recorded only to illustrate the frequency of defective hearing being dependent upon those "adenoid growths" or granulations so admirably described by Dr. Meyer (see *ante*, page 157). The deafness had existed "always;" slight otorrhœa formerly; watch heard only six inches from the right ear, three from the left; tuning fork about equally well on both sides. Tinnitus and "whistling" complained of. Both membranæ tympani indrawn; Eustachian tubes impervious to Valsalva inflation. On using the Politzer bag, air rushed into the right tympanum with a very decided "crack," startling to the patient. The hearing on this side at once became so restored as to enable him to distinguish my watch at twenty five feet. On the left side it had not improved; nor, on again inspecting the membrana tympani of this ear, was its position observed to be changed for the better, while the right one had quite lost its extreme concavity, and appeared to be now naturally placed. On repeating the inflating process, this time with the otoscope in the left meatus, no air entered the drum of this ear. The obstruction, therefore, was probably to be found at the mouth of the tube on that side. Accordingly, my finger, when placed behind the soft palate, detected quite a nest of granulations in the naso-pharyngeal space, close to the orifice of the left Eustachian tube. These were treated for four days with a strong solution of nitrate of silver, which caused them to shrink up. Afterwards there appeared no further resistance to the entrance of air into the tympanum, and on the fourth day the hearing on the left side was as good as on the right. I saw this patient occasionally for a month, and he has attained perfect hearing.

LECTURE XII.

PURULENT AURAL CATARRH, OR OTITIS.

GENTLEMEN,—The inflammations of the middle ear which we have hitherto been considering, are the simple mucus or non-suppurative catarrhs. In the remaining lecture our attention will be confined to the more severe and dangerous aspect of the catarrhal process, which is characterized by the formation of pus instead of mucus. It is fortunately the less common, and would appear to be a higher grade of inflammation, in which there is an excessive development of free cell-formation, leading to what is termed suppuration in the inflamed mucous membrane. Occasionally there is a mingling of both inflammatory products, and so, depending upon which element preponderates, the secretion is called either *muco-purulent* or *puriform mucus*. This particular form of inflammation attends, or rather is the consequence of the exanthemata and of other febrile diseases, as typhus fever, and of phthisis. It also occurs in weakly scrofulous constitutions as the result of slight injuries and inflammations which in a healthy subject would only have caused a simple aural catarrh like those described in the five preceding lectures. Nor must the fact be overlooked, that in persons at all predisposed to puriform affections, a purulent catarrh, leading even to a fatal termination, may be developed from the acute simple form neglected or improperly treated. This formidable disease is much more frequent in children than in adults, and in the former unhappily is often overlooked until a discharge from the ear excites attention, when probably mischief beyond repair, loss of ossicles and perforation, for instance, will have already taken place. The symptoms which are indicative of inflammatory action going on in the infant's ear, although as yet no matter has been discharged (that is, inflammation unaccompanied as yet by otorrhœa), ought to be especially studied.

The diagnosis of infantile aural inflammation is difficult, in comparison with the recognition of the same disease in adults, because young children are not able to explain the situation, kind, or degree of their pain, and it is almost impossible either to make a thorough examination of the parts, or to determine the extent of the deafness. We must not, however, be deterred by these hindrances from making such investigation as will enable us to detect an aural catarrh in an infant when there is as yet no purulent discharge.

Of all specific or contagious disorders to which we are prone during the period of childhood, scarlet fever must be considered the one which most frequently and most destructively affects the organ of hearing. It is unfortunate that the attention of the medical practitioner is in many instances insufficiently or not at all directed to this result of the disease now so prevalent, at the time when direct treatment might arrest or modify its often fatal progress. The excuse sometimes advanced for such neglect, is that in the often almost desperate endeavour to save actual life in this disease, the minor effects of it are for the time disregarded; it being imagined soon enough to attend to the ear when the patient becomes convalescent, or at all events, when he is no longer in danger of death. The foregoing is a fatal error, against which our voices should be earnestly uplifted. Although in cases of great extremity much may not be practicable in the way of local treatment, yet in all cases the aggravating conditions attending this complaint, among which ear disease is one of the most urgent, may be alleviated; and in many, the progress of the latter may be arrested and its severity diminished. Let us not forget that death is not always the most unhappy consequence of a purulent catarrh in an infant's ear caused by scarlet fever or any of the allied zymotic disorders; and that a young life destined to rely for its future welfare or perhaps means of existence upon its own exertions, may be considered as over dearly purchased by total deafness, involving consequent dumbness.

In briefly noticing the principal effects of purulent catarrh upon the ear, it will be convenient to trace them from within outwards, as this is the more usual course of the disease: its *pathology* will thus be constituted. There are two forms of otitis, the *acute* and *chronic*, the latter being most generally a result of the former, though occasionally a chronic purulent catarrh does appear primarily; that is, without any acute stage having preceded it, and pursues a very slow course from the commencement. I shall not, however, as in former lectures when treating of the milder forms of aural catarrh, make a formal separation into acute and chronic; as the transition from one to the other may be too rapid for such distinctions to have any practical importance.

Otitis, or purulent aural catarrh, is manifestly an acute inflammation of the lining membrane of the tympanum, either beginning there or else extending into that cavity from the mouth and throat through the Eustachian tube. The inflammatory action may have been set up by a common cold or in other ways already referred to in Lecture VII., or by some one of the exanthematous disorders, but the effect is that the lining membrane of the tympanum becomes thickened, red and velvety, and it secretes muco-purulent, or as the disease progresses, entirely purulent matter. In some rare cases (even at this stage of the inflammatory process) the secreted matter becomes absorbed, or finds a gradual exit through the Eustachian tubes, leaving the mucous membrane and contents of the cavity more or less disorganized and the auditory functions impaired. In the greater number of instances, however, the secretion becomes more abundant and puriform, and fills and distends the cavity of the tympanum so as to bulge outwards the membrana tympani. It is unable to escape through the Eustachian tubes, by reason of their walls having become extremely thickened, and consequently impervious to air or fluid. Here, and now, it is that this acute otitis, which is precisely what occurs in scarlatina, makes its saddest, its most direful ravages. It may extend its destructive effects in every direction from the tympanum, generally and fortunately outward towards the membrana tympani, but often backwards towards the mastoid cells, inwards to the internal ear or labyrinth, or upwards through the bony walls to the brain. At this point I cannot help expressing my belief that many fatal cases of scarlet fever are in reality deaths from abscess or other disease of the brain continued from the tympanum, but which disease was during life unrecognized by the practitioner. Such a fact has been ascertained by the discharge from the ear being accidentally discovered after the fatal event of the cerebral affection. Facial paralysis sometimes attends purulent catarrh, when the inflammation has extended to the aqueduct of Fallopius, the bony canal in which lies the portio dura nerve. Seldom or never is this peculiar condition (Bell's paralysis it is called) when unaccompanied by otorrhœa, caries or cerebral affection, observed, without some disease existing in the middle ear; and this fact ought to direct the special attention of physicians to the state of the middle ear in every such instance. My teacher, the late Mr.

Pilcher,* has given perhaps the best and truest pathological description of the effects of otitis upon the brain; and as it is also a short one, I will here quote it :—

"The active disease of the mucous membrane of the tympanum, as is too well known to every practitioner, extends its baneful effects to the brain, lighting up inflammation of a character and degree regulated by the peculiar attendant circumstances and terminating as determined, by predispositions natural or recently acquired, by the specific fever. Most frequently the disease take the course of the roof of the tympanum, thus reaching the middle lobe of the cerebrum; or, if a little more backwards, reaching the posterior lobe; but the inflammation, in my experience, usually affects the posterior or anterior part of the brain by extending along the dura mater or the lower surface of the middle lobe, I have met with an abscess in the middle lobe, and a second one, not communicating, in the anterior lobe. The disease also passes through the posterior wall of the tympanum and reaches the cerebellum, and not very infrequently the lateral sinus itself; or it sometimes mounts upwards through or round the margin of the tentorium to the brain proper. I have usually found that the bone is not carious nor necrosed, but inflamed, which inflames the dura mater, which in its turns transmits it to the structure of the brain, and

* On the Effects of Scarlet Fever upon the Ear. A paper read before the Medical Society of London, December 16th, 1854.

there results an independent abscess, *i.e.*, an abscess not communicating with the ear, but surrounded by brain matter; in several instances, I have seen the bone superficially ulcerated under the dura mater which still remained attached ; and in one instance, an abscess in the cerebellum had a small communication with the ulcerated tympanum; in another, the dura mater covering the petrous bone, formed the floor of the abscess, the other walls being in the middle lobe; in another, a direct though small communication existed between the cavity of the tympanum and the lateral sinus." Further extensions of scarlatinous otitis involving the destruction of the *labyrinth*, and where the bulk of the cochlea has escaped from the outer meatus through the tympanum, are described ; and the writer proceeds to say "that he need not add that these desperate cases are necessarily attended by complete deafness, and in young subjects, entail consequent dumbness. Scarlet fever becomes the most fruitful source of deaf-dumbness, far more fruitful even than congenital defects, and perhaps more fruitful than all the other causes combined."

As I myself made most of the *post-mortem* examinations of the above cases and prepared drawings from them, I can personally testify to the exactness of the description. I have also recorded similar ones in my pamphlet on " Diseases of the Ear arising from the Exanthemata," 1853.

SYMPTOMS *and* DIAGNOSIS *of Otitis, or purulent aural catarrh.*

Many of the symptoms and the objective appearances of the membrana tympani in this disease are very similar to those previously described (in the Seventh Lecture) as occurring in a severe form of simple acute catarrhal inflammation. The difference consists in the aggravated character of those now under consideration, and in the more critical condition of the patient. The symptoms now added, may be taken as a continuation of those given on page 110, where it was stated that cases of acute catarrh ending in suppuration would be particularly referred to when the subject of purulent catarrh came under discussion.

Acute otitis is one of the most painful of all ailments. The disease being more common in children than in adults, the symptoms by which it declares itself should be as clearly defined as possible, and strict and special attention paid to them. Parents and teachers have not infrequently boxed a child's ears for "inattention," or "stupidity," when all the while its seeming listless indifference has been caused by the advance of deafness, of which the first beginning was probably in some past attack of distressing pain extending over several days, when neither the parents nor the practitioner they called in understood aright the state of the ear; perhaps they never even referred to it at all.

If the inflammation be not an accompaniment of

one of the exanthematous disorders, the exciting cause will generally be "having taken cold," or exposure to draughts of piercing wind. A sudden paroxysm of pain awakes the child, in the night or towards morning, out of its sleep, and a violent fit of screaming, sometimes attended by convulsions, follows. If this attack occurs at the time of dentition it is attributed to the pressure of the teeth, and perhaps the gums are lanced the next day. Grey powder, castor oil, mustard poultices, and a variety of other treatment and expedients are adopted to relieve the convulsive seizures, which become rather more frequent, and to combat the accompanying fever and delirium. The ears are not examined, although some or most of the symptoms indicative of cerebral irritation are associated with the continued suffering of the child. The very intimate anatomical and sympathetic relations of the middle ear to the brain have been so repeatedly pointed out during these lectures, that I need scarcely here insist upon the absolute necessity that in all cases where cerebral irritation, in infants, is manifested by such signs as are above referred to, the ear should receive just as much of your professional attention as the gums. To proceed with the further description of one of these sad cases. Days and nights are passed in extreme anguish; the tongue is dry and furred, pulse hard and very rapid, skin dry, bowels constipated; considerable fever is also present. The cry of the suffering child is almost unearthly; this is said by some to be peculiar to otitis. Even strong men afflicted with this disease often utter cries or moans of a strange piercing character; and the pain is usually described by them as severe and agonizing. What must it be with young children? The suffering seldom is remitted for any length of time, and the child's snatches of sleep are suddenly broken by paroxysms of stabbing, darting pain through the head and in the ear. The screaming is renewed, the restlessness becomes extreme, the countenance anxious, and the head rolls painfully from side to side, while the baby's interest or notice can no longer be attracted by any object, or the circumstances around it. Convulsions may now set in for the first time, or may recur more often; the sharp cries of agony are succeeded by hoarseness, low moaning, and exhaustion. More decided head symptoms, indicative of an extension of disease to the brain or its membranes (encephalitis or meningitis) succeed this terrible struggle; then coma; and death will presently close the scene of suffering.

Before going on to describe other possible terminations of an attack of acute otitis, it will be useful if I note one or two circumstances which will help you to distinguish this affection, in a child, from some other less severe infantile ailments. The loudness and shrillness of the screaming will show that neither the lungs, trachea, nor other portions of the respiratory apparatus are greatly implicated; because in affections of those organs children never cry loud or continuously. The cry more resembles that uttered by a child suffering from inflammation of the bowels, or a primary affection of the brain, but the other symptoms indicative of such disease will be absent from these cases. The circumstances under which the pain appears to be increased or mitigated, and how, should be carefully noted; for, owing to the inability of children to designate the situation of their pain, and the difficulties always in the way of making a sufficient examination, we have to attend to every minute particular if we care to make our diagnosis really certain. When the middle ear is the seat of disease, the pain will be increased by every rough movement of the body; the child will cry more and more every time it is "jumped" in the arms of the nurse, or in any way violently shaken; changes in the position of the head call forth renewed expressions of pain, and if only one ear be affected, the screams will be more distressing when the child is laid on that side; any loud noises disturb; cold air and draughts, and also bright light, are shrunk from; and every effort at swallowing, sneezing, coughing, and the usual accompaniments of a common cold in the head will all aggravate the pain. If being suckled, the infant will move itself away from the breast, or push aside the bottle after a few attempts to suck, while its usual nourishment will be taken somewhat more easily if administered with a spoon. Warm water poured frequently into the ear, hot soft applications, gentle bathing or steaming, and perfect quiet, will always soothe the pain. In other disorders with which this affection might on a mere cursory examination be confounded, none of the above-named actions or personal movements would have the effect of aggravating the pain, nor would the palliatives mentioned give sensible relief. Otitis in the youth or adult is attended by symptoms of like character to those just described as occurring in the infant; but for obvious reasons it will not be so difficult of recognition. It need scarcely be said that the consequence, whether in a child or grown person, is always impairment of hearing, and sometimes total deafness ensues.

The objective appearances, or *local* symptoms in the ear, are likewise at their commencement very similar to those of a severe case of simple acute catarrh (page 111). The inflammatory action having commenced from within the tympanic cavity, the membrana tympani will not at first have lost its transparency; and the congested thickened mucous membrane, with its distended blood-vessels, may be at the beginning distinctly seen through the drumhead. In other instances, opaque fluid may be distinguished, lying within the cavity. As the disease makes progress, the membrana tympani becomes flatter and duller, darkish red in colour, and streaked with blood-vessels on the outside. The collection of pus behind presently bulges it outwards; its external surface has a sodden look, and it is sometimes so covered with macerated epithelial scales which have peeled off from the swollen and thickened lining of the external auditory passage, that it can scarcely be seen at all. The auricle itself, especially about the tragus and concha, participates in the inflammatory action, and is œdematous, swelled, and shining, or occasionally of a livid colour, hot, and tender to the touch. In later stages of the disease, when it has continued for some days, the skin over the mastoid process also becomes reddened and tumefied, and painful when pressed upon. This swelling behind the ear is serious in character, and dangerous if not properly attended to. It therefore requires to be examined with great care; and considerable but cautious pressure must be made with two fingers upon the soft swollen and œdematous integument, in order to detect, by the degree of fluctuation or "pitting," the possible presence of matter, which may be deeply seated beneath the fascia. If the patient's throat be looked into, the pharyngeal and faucial mucous membrane will appear swelled, congested, or granular, and the Eustachian tubes are probably rendered impervious by extension of the disease; the general condition of the throat resembling that observed in scarlatinal inflammation.

Some very exceptional cases do now and then occur, where an abscess in the tympanic cavity forms so very gradually, that it will run its course and cause a perforation of the membrana tympani without producing severe pain or any great constitutional disturbance.

The immediate *terminations* of a purulent catarrh have been admirably described by Wilde.* I will here briefly enumerate them. 1st. *Resolution*, as already referred to, in which the pain gradually subsides, the swelling lessens, and the hearing is restored, although tinnitus may persist for a long time.

---

* Wilde's "Aural Surgery," page 332, and this description has been closely imitated by Von Tröltsch "On the Ear," page 378.

Strictly speaking, under this heading (of resolution) must not be included hypertrophy of the mucous membrane of the tympanic cavity, nor the deposition of consolidated purulent or muco-purulent effusions, which may originate severe deafness with constant "noises" (see Case 2 in the preceding lecture).

The 2nd and most common termination is, that the pent-up matter bursts suddenly through the membrana tympani as the readiest external outlet, and is discharged; *otorrhœa* is thus established.

The 3rd termination of otitis, by extension of the inflammatory process to the brain and its membranes, is always dangerous, and often fatal. The various routes by which a purulent catarrh of the tympanum may advance inwards towards the brain have been specified already when giving the pathology of otitis (page 227). I shall have occasion, in speaking of the treatment, to refer again to the head affections which accompany otorrhœa or are consequent upon it, such as *cerebral abscess*, and *caries of the temporal bone*; but these being happily not frequently met with as resulting from "aural catarrh, or the commonest forms of deafness," will require no lengthened description in these lectures. To proceed with what we have called the second termination or development of otitis,—OTORRHŒA.

I do not think that the membrana tympani is often perforated as the result of inflammation confined to its own proper laminæ; but its structure becomes softened by the inflammatory action proceeding from within, and is then more easily ruptured. This commonly happens in acute otitis, as it occurs in the exanthemata, in severe naso-pharyngeal catarrh, or primarily also during other disorders. In other cases, a sudden excessive pressure of air in the cavity of the tympanum leads to rupture of the membrane, yet *only* where the mucous lining of the tympanic cavity is already diseased, and the proper tissue of the membrana tympani has suffered the change above indicated. We may observe this process not unfrequently in cases of purulent catarrh affecting cachectic, tuberculous, and scrofulous patients, and when it has commenced with scarcely perceptible inflammatory symptoms (perhaps with only slight tinnitus aurium, or occasional pricking sensations in ear), or even without any noticeable subjective symptoms at all. This we may term *chronic otorrhœa*, inasmuch as the beginning of it may be traced back to early childhood. Pain was only felt when some of the exciting causes had been at work, followed by ulceration; or perhaps the first intimation that the patient has had of any aural complaint has been a whistling of air through the drum-head after blowing the nose or sneezing, followed at once by a more or less copious discharge from the ear.

The ceruminous secretion in the meatus having before entirely ceased, the cuticular lining becomes detached in whitish flakes, and mixes with the discharge from the tympanum. If we syringe away the discharge, the exposed surfaces of the drum-head and the meatus are seen to be villous, red, and vascular; all traces of cuticle have disappeared, and the whole auditory canal is diminished in size, contracted, and converted into a muco-purulent secreting surface.

After careful removal of the secretion, and cleansing the meatus, we may find the aperture in any part of the membrana tympani, and of any size, from that of a pin-hole to complete absence of the membrane. According to Wilde, the opening forms more frequently in the anterior part of the membrane, in front of the manubrium of the malleus, the reason being in his opinion because this part being opposite the entrance of the Eustachian tube, it is more exposed to the shock of air striking against it from within. I cannot say from my own observation that rupture more generally takes place at this point than at the posterior portion of the membrane,—between the malleus-handle and the periphery. Besides this, the Eustachian tube is often closed in acute purulent catarrh, in which case the membrane would not be more influenced at one part than another, by air-pressure from within. The reason why rupture occurs more frequently at the parts lying between the manubrium and the cartilaginous ring, whether in front of, or behind the former, is to be found, as I think, in the fact of the membrane being weaker and thinner there. There are, however, many cases in which perforations occur at the lower, or the upper portion. I have several times seen two apertures in the same membrane, one behind the malleus, and the other in front of it.

In general the size of the hole is about that of a

small pea at first, but the breach enlarges by gradual ulceration of its edges. The extent of the loss of substance bears no proportion to the quantity of matter discharged, nor to the duration of otorrhœa. The largest perforations occur after scarlatina, the tympanic cavity becoming quite exposed and open to the external air. Occasionally some of the ossicula auditûs become disconnected and are discharged; this happens especially with the incus, which is less firmly fixed than the others. When very considerable or entire loss of the membrana tympani has taken place, we are able to distinctly see the mucous membrane lining the inner wall of the drum; and sometimes, if the swelling is not too great, the position of the ossicles may be ascertained. Even the anterior border of the entrance into the fenestra rotunda is occasionally visible, but the *membrane* of this fenestral opening cannot be distinguished, because of its oblique, and fortunately, secure and protected situation in a niche. If the perforation is in the upper or posterior part of the membrane, the long process or the whole of the incus are frequently wanting, and the connection thus broken between the stapes and the other bones comprising the chain. It is important to note that this solution of continuity between the ossicula auditûs, especially in the articulation between the stapes and incus, may take place in the purulent inflammatory process, without necessarily any rupture of the membrana tympani; and it is in such instances as these, that a most surprising improvement in hearing follows when Yearsley's "artificial tympanum" is applied to the membrane; the incus being thus pressed against the stapes, and the continuity of the chain restored.

Occasionally, when the incus is wanting, we discern the minute head of the stapes lying at the posterior and upper edge of the invisible labyrinth wall; it will look like a small elevation covered with reddened mucous membrane.

The appearances within are very varied indeed, and change during the progress of disease or treatment; the principal alterations depending upon the condition of the mucous membrane on the promontory, which is sometimes swollen, uneven, or granular, or projecting in a soft, reddened mass outwards, so as to prevent any other of the contents of the tympanum from being seen.

The membrana tympani (or as much of it as remains after the perforation) is thickened throughout all its layers, and dull and infiltrated in appearance if not partially calcified and its surface covered with secretion. Some portions may be adherent to the walls of the tympanum, especially to the promontory. The curvature is often altered, and the edges of the perforation may be either bright red, or of a dirty-grey colour; they are jagged and variously shaped, and covered with a creamy fluid. Sometimes the malleus-handle is denuded of its covering, and lies exposed in the gap, while its head and short process appear clearly defined at the upper wall of the meatus, close to the periphery of the membrane. In all cases where the hole is situated about the centre of the membrane (the *umbo*, or most concave part), the malleus-handle from having lost its proper attachment to the membrana tympani, falls inwards deeper into the tympanic cavity.*

Diagnosis of perforation.—It is not generally difficult to ascertain the existence of an aperture in the membrana tympani. While the meatus is being cleansed by careful syringing with warm water, the fluid injected will be observed to return with a yellow or yellowish-green colour, and to have floating about in it some greyish flocculi, which are not dissolved. The former is pus, the latter mucus. Sometimes the first will predominate, but the presence of the flocculent mucous secretion is at once decisive of the existence of an opening into the tympanum, because there only could it have been formed. If the water is felt trickling down the throat by the patient, or if the fœtid discharge, thinned by being mingled with the water, causes a nauseous taste in the mouth, this will be conclusive evidence that some of the water has passed through an aperture in the membrane into the tympanic cavity, and down the Eustachian tube to the throat. Bubbles of air, or the pulsating movement (synchronous with the heart's

---

* It is impossible to describe here at greater length the manifold appearances presented in perforations of the membrana tympani. They will vary in accordance with the size and situation of the rupture, and the condition of the parts inside the tympanum; but the student who desires further information on this subject will be well repaid for reading Politzer " On the Membrana Tympani." The description of the appearances in perforation given by that author is most faithful, painstaking, and elaborate.

action), of a drop or two of fluid at the bottom of the meatus, would lead to the suspicion of an aperture. This diagnostic sign was first described by Wilde. I have, however, a case now under treatment, in which the pulsation of a few drops of purulent secretion on the floor of the auditory canal is always visible, yet certainly there exists no perforation. I am inclined to suspect that in this child the carotid canal is imperfect in its bony parietes. If, when the patient blows his nose rather violently, or presses air by the Valsalva method into the drum, a whistling sound is heard through the otoscope, a small perforation may be diagnosed: or, should the Eustachian tube be at this time impervious, the better course is to cause air to pass through by the Politzer mode of inflation, when, if a hole exists, we shall hear the usual characteristic "perforation-murmur" (see pages 80 and 94). At the same moment, perhaps, some of the secretion will be forced into the meatus. An aperture, even of considerable size, not unfrequently exists without otorrhœa; but upon the slightest accession of cold, puriform mucus will be secreted from the surface of the exposed tympanal cavity, and otorrhœa will ensue.

The impairment of function.—*Deafness*, resulting from perforation of the membrane, bears no proportion to the loss of structure. With small holes we may find almost total deafness; while, on the other hand, patients with perforations of a moderate size, or even large ones, are sometimes considered by their friends to be scarcely hard of hearing. It is wonderful how the loss of substance may be compensated for by the use of the "artificial tympanums," combined with appropriate treatment. I have patients now under observation (cases will be subjoined) in whom the whole membrane has been lost, together with the malleus and incus: others with very large perforations; who all can distinguish the ticking of my watch across the room. This would seem to show that the deafness in these cases depends mainly upon the condition of the drum-cavity, and especially upon that part of it where the fenestra ovalis and fenestra rotunda are situated. We have repeatedly seen in the course of these lectures how importantly the hearing is affected by pathological changes at these two fenestræ; so that a very slight interference with their functions may cause severe deafness. But if the mobility of the stapes in the fenestra ovalis is not much diminished, nor the membrane of the fenestra rotunda thickened, waves of sound can still pass through the perforation of the membrana tympani across the cavity and reach the labyrinth. I am convinced, indeed, that if due support or pressure can be given to the stapes, in cases where the two other ossicles and the membrana tympani are lost, vibrations may, through the fenestra rotunda *alone*, reach the labyrinth fluid and the nervous expansion, in number and intensity sufficient to produce excellent audition. The following case, I think, proves and exemplifies what I mean :—

A young lady is at present under occasional treatment for remaining deafness on the right side, from whose left ear the tympanal membrane, malleus, and incus, are all entirely gone (in consequence of scarlet fever nine years ago); yet, in the same ear I can render the sounds of my metronome higher or lower, loud or dull, to her perception, solely by regulating the pressure upon the stapes by means of the little cotton plug. Here, assuredly, the waves of sound do not pass *into* the wool and stapes, and so arrive at the perilymph; but they pass directly to the membrane of the fenestra rotunda, which is made more or less tense by the pressure at the other end of the labyrinth (at the fenestra ovalis), and in this way impress the labyrinth fluid and the auditory nerve.

These remarks are made incidentally for the purpose of showing that the generally received opinions on this branch of physiological acoustics need to be modified.

AURAL POLYPI, Etc.

These growths sometimes, but very rarely, become developed from the surface of the membrana tympani or meatus, in the advanced stage of simple chronic catarrhal inflammation, without any purulent discharge co-existing. The formation of polypi is, however, one of the commonest results of long continued otorrhœa. They may be described as vascular tumours, vegetations, granulations, and fungoid growths; but their classification under different names according to structure or size is of no practical importance. In connection with purulent aural catarrh the above terms may be used indifferently. Polypi or granulations, then, may be said to arise in the course of chronic otorrhœa with perforation of the membrana tympani; and may usually be detected by a thorough and careful examination, lying at the bottom of the meatus, imbedded in purulent secretion. They vary exceedingly in size, shape, density, and position, being in some instances so small as to escape detection, or to appear as excrescences at the bottom of the meatus; while in others they are of such dimensions as to fill the external auditory passage and project beyond it. You may have recently seen me remove one of such large size as to have caused by its gradually increasing pressure, such an enlargement in the calibre of the meatus, as to allow of my little finger being passed half an inch into it.

When a perforation exists, the polypus usually springs from the mucous membrane of the tympanum, very often just behind the membrana tympani; and sometimes one will originate from the upper portion of the Eustachian tube. It then passes out of the cavity through the aperture, and, spreading out over the external surface of the membrana tympani, may give the surgeon the impression that he is looking upon the exposed and open tympanic cavity itself; or it may be erroneously imagined that the polypus has grown from the surface of the drum-head, or from the walls of the meatus contiguous to it. Toynbee has treated of polypi as belonging to the meatus; but the inefficient method of illuminating the auditory passage in his time, will probably account for his mistake in this respect. Error as to their origin may generally be avoided by passing a small blunt-pointed probe bent at right angles, cautiously round the margin of the growth, which can, if it proceed from the tympanum, be lifted off the external membrane. There may be only one polypus, or two, even three, may be lying side by side, growing from different parts, *i.e.*, one may proceed from the tympanum, while another may originate from the meatus, the surface of which has become granular and softened; but examination and inspection will enable us to distinguish whether the growth arises from the middle or the external ear. When situated at the bottom of the meatus, and small in size, polypi are generally of a bright red colour, fleshy, pedunculated, and soft in texture, but when grown so large as to appear near the outlet of the meatus or project into the concha, they become pale, cuticular on their surface, and comparatively insensible to touch except near their root. The smaller, raspberry-shaped or lobular ones are more vascular, friable, and gelatinous, and will readily bleed when touched or probed. The structure of all these formations is complex, and varies considerably, but it will usually be found to correspond very much with the tissues from the surface of which they spring. Polypi are very liable to recur after they have apparently been destroyed.

TREATMENT *of Purulent Aural Catarrh, or Otitis.*

The treatment of an acute otitis must be decidedly antiphlogistic, and conducted with energy. The measures previously recommended (page 121) in simple acute catarrh must all be carried out to the fullest extent; for we have now to deal with a more severe form of disease. The patient must be strictly kept to a warm, well-ventilated apartment, or when the symptoms are more developed and threatening, confinement to bed is absolutely necessary. General bloodletting will be inadvisable, for the same reasons as before assigned, but leeches must be applied as often and in such number as the urgency of the attack requires, and having regard to the general condition of the patient. I know no class of cases in which local depletion is so imperatively needed or in which relief from agonizing pain is so speedily afforded by it as in aural inflammations. As when speaking of the *symptoms* of otitis, I requested that what I then said might be considered as supplementary to the description which had been given before, so with reference to *treatment*, any details now added may be taken as a continuation of those which before occupied us during half a lecture (pages 121 to 130). Assuage the well-nigh intolerable pain of acute otitis by all possible means, local or general, excepting always those which are contra-indicated in cerebral affections; because our chief aim is to check the onward progress of this most insidious disease to the brain or its membranes. Even in this complaint it is not, in my opinion, advisable to have recourse to mercury in any form, as a medicine. I believe the use of it will neither arrest the aural inflammation nor relieve the symptoms of cerebral irritation, while it is well known that when suppuration has once commenced in the brain the effect of mercury is most injurious. The bowels should, in this as in all febrile affections, be freely acted upon. The skin being generally hot and dry, diaphoretics combined with sedatives are required in the early stages

Abstinence from animal food must be insisted on, and warm baths for the feet, or the whole body, with all other means calculated to allay the inflammatory fever, should be had recourse to.

For children, the treatment must of course be proportioned to their more tender age and feebler powers; and more frequent examination of the ear will probably be necessary with them, because they cannot express their sufferings and wants in words. The medical attendant ought to be constantly on the watch for every indication of a suppurative process being set up in the ear. Remember that vigilance and prompt attention to the condition of the latter, in the acute infantile aural catarrh so often attending or following scarlet fever and the allied disorders, may not only save many an infant's life, but prevent it from growing up a deaf-mute. Both in the child and the adult, frequently filling the ear with warm water, medicated with hyoscyamus, opium, &c., will mitigate the pain, and also encourage suppuration when the process is considerably advanced, and we desire that the matter should be evacuated through the membrana tympani. Poppyhead fomentations, steaming, linseed meal poultices, or the new "epithema," are all useful for the same purpose. Injections up the nostrils should be employed to remove muco-purulent secretion; and where, as in scarlet fever, the otitis is accompanied by diffuse inflammation of the throat, pharynx, tonsils, &c., the greatest possible attention should be paid to it, because this condition (of the throat) has in all probability caused the aural disease. Cleanse the throat as well as the nose by frequent use of "spray" injections with such syringes as are shown on page 200, and prevent as much as possible the entrance of vitiated secretions into the respiratory organs, or the Eustachian tubes along the continued mucous membrane.

It is obviously most desirable and important in this class of cases (the scarlatinal) to secure, if we are able, the evacuation of the pus which is collecting in the cavity of the tympanum, by its natural outlet through the Eustachian tube. To this end it is well to employ the Politzer bag at an earlier period than I indicated to you when speaking of the treatment of simple aural catarrh. It must be used with as little force as possible, and we have already seen how easily inflation is accomplished in children, without even requiring them to swallow at the moment of the bag being squeezed. I recommend this gentle method of opening the tube in preference to the use of the catheter, even on grown persons, because, by the latter instrument, inflammation of the nasal cavities and faucial mucous membrane is often increased. In some few cases the purulent accumulation in the tympanum and aural extremity of the Eustachian tube will flow out of the canal when rendered pervious, thus bringing about the "1st termination" of the disease, or *resolution*, spoken of some pages further back.

When this does not occur, and we are assured, by the bulging of the membrana tympani, and by other objective symptoms, that pus has accumulated in the cavity (especially if there be delirium, convulsions, or any other alarming signs of cerebral irritation,) incision, or puncture of the membrane ought to be performed without delay. The effect of the operation under these circumstances is as immediately and strikingly beneficial as that of opening *any* abscess surrounded by dense or bony structures. The distressing pain is at once relieved, and the extension of the disease inwards to the brain, or temporal bone, is as far as possible prevented. The matter will now pour out through the artificial opening, and the case becomes similar in most respects to one where the pus is evacuated spontaneously, which has lately been described as the "2nd termination" of this disease, or "perforation." It should be noted, however, that when the membrana tympani is opened by incision, the consequent loss of structure is slighter than when it becomes perforated by nature in the suppurative process and in ulceration. The mode of performing the incision is very simple. I make use

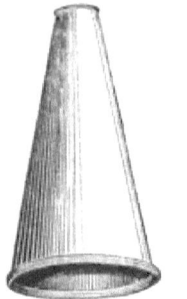 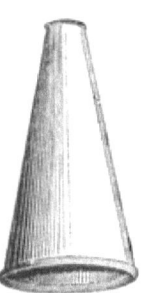

GRUBER'S SPECULUM (TWO SIZES).

of a small double-edged cataract knife, bent at a right angle, and inserted into a handle. I introduce this instrument through Gruber's speculum, while a good light is thrown to the bottom of the meatus from a concave mirror attached to my forehead, thus leaving both hands at liberty. Some portion of the membrane will generally be observed to be more distended than the rest, or bulging outwards. This spot, wherever it may be, should always be selected for the incision. It will usually be near the middle, or on one or other side of the malleus; because at this part the fibrous laminæ (substantia propria) being thinner, offer less resistance to the pressure or ulceration from within. If there is no decided appearance of flattening or pushing out of the membrane to be observed on inspection, and yet unequivocal symptoms are present of matter having formed in the middle ear (perhaps filling principally the mastoid cells, and therefore not to be detected on inspection), I have in three or four instances employed the Politzer bag, and *forcibly* ruptured the membrana tympani from within. This mode of relieving the continuous suffering where the pain and other subjective signs (together with the progress and history of the case), clearly indicate that otitis has reached the suppurative stage, has not, so far as I know, been before adopted by any aural surgeon since the discovery of the Politzer method of inflation; but the proceeding has been entirely successful, and without causing any shock to the system. The portion of the membrane thus perforated artificially from within has always been the thinnest and weakest. In the cases where I have had an opportunity of watching the progress of *healing* of the perforation made in this novel way, I have always noticed that the otorrhœal discharge diminished gradually before the opening became closed; and I am therefore confirmed in my opinion, that so long as a purulent secretion in the tympanic cavity continues, no permanent cicatrization of the opening—whether the latter have been artificially made or not—is possible. This new procedure, of opening the membrane by a strong air-douche when circumstances indicate the feasibility of such an operation, seems, I think, to offer some decided advantages over that of external incision. It drives the collected matter outwards at once, and is, moreover, a close imitation of the manner in which a rupture of the membrane happens naturally when the state is favourable for it—by the patient's sneezing, or forcibly blowing the nose.

If, in the course of the disease, the *mastoid process* has become tender to pressure, the integument over it red and tumefied, and particularly if there is any (even indistinct) sense of fluctuation to be discerned, prompt measures are required to prevent caries of the bone, or mischief extending to the cranium and its contents. We must not hesitate to make an incision into the affected parts *down to the bone.* The periosteum must be freely divided, to liberate the pus confined beneath, and to relieve the painful tension. The swollen, infiltrated, and doughy condition of the structures covering the mastoid process, often renders it necessary to introduce the knife nearly an inch deep into them, and sometimes the incision requires to extend the whole length of the process. To avoid the posterior auricular artery, the cut should be made from half to three-quarters of an inch behind the auricle, but parallel with it. Great relief immediately follows; usually there is an escape of pus, but sometimes it is too deeply situated in the cells. The hæmorrhage from the operation may be considerable, but is easily arrested by a dossil of lint. The cut surfaces look brawny, as in phlegmonous erysipelas. The management of a case after incision will mainly depend upon the circumstances attending exfoliation of bone, &c. Poultices should be applied for the purpose of encouraging the discharge and the separation of dead bone, if there is any to come away. If not the wound will shortly heal.

Such promptness in dealing with mastoidal abscess is imperatively demanded. The disease is highly dangerous, and, if recourse is not had to incision, generally fatal. If the pus, in consequence of its deep situation in the cells, has not been reached, and signs of cerebral irritation, such as giddiness, vomiting, pains in the head, &c., are still persistent, we must endeavour to evacuate the purulent collection by means of perforating the mastoid process with the trephine. This operation is a simple one, and has generally been successful when not deferred too late.

TREATMENT *of Perforations, Otorrhœa, Polypi, &c.*

The most desirable result to be obtained by treatment in a case of perforation is the restoration of the mucous membrane of the tympanum to a healthy state, and then closure of the orifice in the membrana tympani. If we attempt to lessen or close the opening before the purulent discharge is checked, and the condition of the tympanic cavity improved, we shall probably increase the disease and deafness, instead of curing both; because the exit of matter will be rendered more difficult, or hindered altogether; and all the evil consequences of retained fœtid pus will be induced. It is not so much the hole in the membrane that impairs the hearing as the collection of masses of inspissated offensive discharges on the parts which conduct sound—such as the ossicles and the mucous membrane of the fenestral openings into the labyrinth. Hence the great variableness of the hearing power in these cases. It should not be forgotten when a perforation exists, that such morbid viscid secretion as is formed in purulent catarrh is quite sufficient of itself to maintain and aggravate inflammatory action; for if it be syringed away we can observe the lining membrane of the tympanum to be thickened, red, velvety, and granular, or perhaps containing vegetations or small polypi. The recommendation of many authors, that the healing of a perforation should first be sought, by touching its edges with nitrate of silver, is, therefore, in my judgment, not a safe one. Our treatment must from the beginning be directed to the cure of the suppurative inflammation,—the *otitis*, and its result, the *otorrhœa*.

Thorough removal of the secretion, by careful syringing with warm water, is essentially necessary, and should be done two or three times a day. If the cavity is much exposed, a little potash or soda may be added to the water. With the greatest precaution and the gentlest syringing, either vertigo, fainting, or sickness will perhaps ensue, but to lessen these annoyances or avoid them as much as possible, you always see me use a small syringe and inject a minute stream. When the perforation is small, removal of the morbid secretion cannot be perfectly effected by injections through the meatus, for in these cases doubtless the air spaces in the mastoid cells as well as the recesses in the tympanum are crammed with it. An attempt should therefore be made by the patient to dislodge or force the accumulations from within outwards by simply practising the Valsalva experiment of inflation, before the syringing is commenced. When this cannot be effected, I use my form of Politzer bag, which easily opens the Eustachian tubes, and masses of morbid secretion may sometimes be seen extruded from the cavity into the meatus afterwards. In the early stages of treatment, I disapprove of using "very considerable force" in injecting strong alkaline ("saturated") solutions into the tympanum by means of a syringe fitted tight to the meatus with india-rubber tubing over the nozzle. This procedure has been advised, but I have on more than one occasion heard it complained of by patients as severe and injurious, and I believe such complaints to be well-founded.

The longer an otorrhœal discharge continues in the ear, the more offensive and disgusting will be the smell arising from it. "Some persons" (I quote from Wilde) "are content to go on through life with this filthy loathsome disease about them." "Some may endeavour to conceal this affection, others are deterred from taking proper advice by the prejudices of their friends, and even of their family medical attendants; they are afraid of the effects of healing or drying up the discharge from their ears!" I, like Sir W. Wilde, have never been able to discover one well authenticated instance where disease in the head has supervened as a consequence of checking otorrhœa, in a case where the condition of the ear had been previously with due care ascertained, and no caries had before existed in the temporal bone. "Men do not distinguish between the *post hoc* and *propter hoc;* and mixing up cause and effect, regard a symptom as a disease." The *prognosis* of otorrhœa, and the *morbid changes* which long-neglected aural changes may lead to, are subjects which will not be dwelt upon in any detail here; but I must warn you to consider that they are of vital importance in your treatment of all these diseases. Be cautious, on all such points, in your replies to the eager questions of patients, for the reason well assigned by Wilde, that "*so long as otorrhœa is present, we never can tell how, when, or where it may end, or what it may lead to.*" "For this very cause, if no other or better existed, the old doctrine of 'letting alone,' or 'leaving to nature,' such affections should be exploded, and we should by every means in our power endeavour to heal them."

There are not many cases of otorrhœa attended with perforated drum-head, which will certainly yield to simple injections of warm water, so that it will probably be necessary to act by means of astringents upon the tissues which furnish the morbid products. The ordinary practice of distilling a few "Ear drops" into the meatus is of little service. The lotion selected should be poured into the ear, night and morning after syringing, until it fills the auditory canal; and should remain in there from five to ten minutes,—the head being kept resting on the opposite side,—and then allowed to run out. In order to bring the astringent lotion freely into contact with the mucous membrane of the cavity of the tympanum, it is necessary from time to time to force a current of air *into* it through the Eustachian tube; for, recollect, if the purulent secretion is retained and becomes hardened and calcified, it may cause rigidity of the ossicles, impaction of the stapes, and consequent severe deafness. This peculiar inflation may be effected by the patient himself, in the following manner. While he sits or lies with the head properly inclined to the one side, and the meatus is filled with the lukewarm lotion, he is directed to inflate by the Valsalva method; or air is forced into the tympanum by using the Politzer bag in the way so frequently described. The air which passes through the tympanic cavity, enters the auditory canal now filled with the lotion, in the form of bubbles; and simultaneously with its displacement of the fluid, the latter will enter the cavity through the perforation. This often takes place so rapidly that the lotion runs down the Eustachian tube into the throat during the operation. Patients who are able to inflate the tympanum themselves by forcibly blowing in air can of course practise this treatment daily without assistance; but it is better for children or timid females to attend the surgeon in order to have it done for them by so easy and simple a method as that of the modified Politzer bag.

Avoid the use of lotions which may form precipitates with the otorrhœal discharge, such for example, as the acetates of lead, and perchloride of iron. The various preparations of zinc stand first in their usefulness for all purposes. If you observe my prescriptions in the numerous cases of offensive otorrhœa which present themselves for treatment at this hospital, you see very generally the following :

℞ Zinci sulphatis,
  Acidi carbolici, āā gr. iij. ad v. ;
  Aquæ rosæ, ʒj. Misce ; fiat lotio.

That most efficient deodorizer, carbolic acid, may be combined with nearly all the astringents. So effectual is the above lotion in bringing about a cessation of the discharge, and destroying its disgusting smell, that I seldom have occasion to vary the prescription. I think it therefore needless to lengthen the list of agents which may be used to produce this desired effect. Small quantities of powdered alum, blown into the bottom of the meatus through a glass tube, will sometimes speedily diminish secretion, and induce shrivelling of the granulations. It causes little irritation, and boils in the meatus do not occur so often when pulverized alum is used, as when solutions of it are employed instead. The foregoing are the astringents which I myself prefer; still, not to restrict you dogmatically to my own practice, I should inform you that " strong solutions of nitrate of silver (even 20 to 80 grains to the ounce) are said on the authority of

Dr. Schwartze, to do very excellent service. These solutions are introduced into the meatus through a glass tube, and blown through the Eustachian tube so as to reach the fauces. Immediately afterwards the ear is syringed out with a solution of common salt, causing it to pass as far inwards as the caustic has done. The insoluble chloride of silver is formed and comes away in large flakes." But what if the neutralization of this strong caustic should not be effected? and we can scarcely imagine that each subsequent injection of salt and water would always proceed in the same track as the former one. You will do well to abstain from such severe measures. Mr. Hinton recommends powdered talc (French chalk) to be blown into the meatus when the ear has been well dried after syringing. "I know," he writes, " of no more effective plan than this, but it is somewhat tedious, and requires daily attendance, and is therefore naturally reserved for the more obstinate cases." I have had as yet no experience of this treatment, and own that I feel disinclined to try it, chiefly on account of its acknowledged tediousness, and, as I conceive, aptness to increase, rather than to diminish, deposition in the cavity of the tympanum. Other remedies can be had recourse to, more expeditious and certain, less painful and objectionable in their nature.

Vegetations, or small polypi growing from any visible part, whether in or out of the cavitas tympani, must, on account of their tendency to maintain the otorrhœa, be removed (when they can be reached), by mechanical means, or else be destroyed by touching with perchloride of iron, or by solid nitrate of silver. I generally use the latter, melted and made to adhere to a piece of silver wire bent at right angles. The growths, and no other portions of tissue, are then to be touched with the point on which the caustic has hardened.

Large polypi must be removed by suitable instruments. Wilde's "snare or noose," or Toynbee's "lever-ring forceps" are usually recommended, but both are rather complicated instruments, and neither of them act, in my judgment, so efficiently as properly constructed forceps. I use a slender three-bladed pair of forceps, which, when open, slide over the polypoid growth to the bottom of the meatus. The polypus is then grasped and held by the little "tiger-teeth" at the end of the blades, as soon as compression of the handle of the instrument is made. (This handle, it may be noted, is placed at right angles to the blades, in the same manner as with all my instruments, to enable the surgeon to see what he is doing.) A twisting motion while the forceps are withdrawn will bring out the polypus. Further treatment is afterwards needed, to destroy the root of the polypus, and to prevent its reproduction; for as soon as one of these troublesome growths is removed, it will begin to sprout again, thus rendering their complete extirpation a matter of time and perseverance. I generally attain this object by caustic applied in the limited manner above described for the removal of granulations. If the base of the polypus has been large, I insert into it the blunt point of a probe, previously dipped in solution of chloride of antimony, zinc, or iron, immediately afterwards syringing away the *débris* of the destroyed morbid tissue. To hasten the conclusion of a case, I repeat this procedure two or three times at a sitting. In this way pain is prevented or reduced to a minimum. As in the treatment of granulations, nitrate of silver is an excellent application in the later stages. Undiluted liquor plumbi may be substituted for the caustics when but little of the polypus remains.

Constitutional treatment.—If we have reason to believe from the appearance of the patient, and from the history of the case, that a generally deranged state of health contributes to keep up the local disease, we should by every means endeavour to improve the tone of the system, by tonics, sea-bathing, change of climate, &c. But here, as in other forms of aural disease, the constitutional treatment must be reckoned as subsidiary to the local. If our cold winter climate and north-east winds excite an aural discharge in patients predisposed to otorrhœa, a residence during winter in a warmer country is clearly advisable when it can be accomplished; but local remedies, such as are suggested above and in the section to follow will, in persons otherwise healthy, be quite sufficient to effect a cure of this most disagreeable and offensive result of purulent aural catarrh.

"ON A NEW MODE OF TREATING DEAFNESS WHEN ATTENDED BY PARTIAL OR ENTIRE LOSS OF THE MEMBRANA TYMPANI, ASSOCIATED OR NOT WITH DISCHARGE FROM THE EAR."

Such was the title of a paper published by the late

Dr. Yearsley in the *Lancet*, July 1st, 1848, and in subsequent numbers. But let the author* of this discovery relate the incident which led to the adoption of what he termed the "artificial tympanum."† He prefaces the subject by saying: "Up to the present time no successful mode of treating perforations of the membrane of the drum, or the relief of the accompanying deafness has been discovered by the profession at large. The only means resorted to has been the removal of pus, or mucus, from the tympanal cavity by syringing, or rendering it free by passing air through the perforation, by way of the Eustachian tube. Either of these proceedings will produce a temporary improvement of the hearing in cases where the tympanum suffers from obstruction, but in many others, when such a state does not obtain, they are of little or of any service. I have, however, the extreme gratification of promulgating a mode of relief for deafness attended by loss of the membrana tympani, which will cause great surprise among the readers of the *Lancet*, not less from its extreme simplicity than from the extraordinary success which generally attends its employment. In 1841 a gentleman came from New York to consult me under the following circumstances: He had been deaf from an early age, and on examination I found

---

* Yearsley "On Deafness," page 219.

† The membrana tympani is in ordinary phraseology called (erroneously, however) the drum or tympanum, but this expression is anatomically wrong, the membrana tympani being the drum*head*, and the tympanum the *drum*. (See page 20.)

great disorganization of the drum of each ear. On my remarking this to him, he replied, 'How is it, then, that by the most simple means I can produce in the left ear a degree of hearing quite sufficient for all ordinary purposes; in fact, so satisfied am I with the improved hearing which I can myself produce, that I only desire your assistance on behalf of the other ear.' Struck by his remark, I again made a careful examination of each ear, and observing their respective conditions, I begged him to show me what he did to that ear, which I should have unhesitatingly pronounced beyond the reach of remedial art. I was at once initiated into the mystery, which consisted of the insertion of a spill of paper, previously moistened at its extremity with saliva, which he introduced to the bottom of the passage, the effect of which, he said was 'to open the ear to a great increase of hearing.' This improvement would sometimes continue an hour, a day, or even a week, without requiring a repetition of the manipulation." Dr. Yearsley then proceeds to relate a case, not only because it was the first which he relieved by his novel plan (by another form of artificial tympanum), "but because he was in a position to show the permanency of the remedy;" for the young lady who was thus successfully treated had been for nearly five years "restored to society, from which, by reason of her severe deafness, she had been almost entirely excluded." She herself writes, "When the aid (the artificial tympanum) is removed I scarcely hear at all."

This new remedy consisted of a small pellet of moistened cotton-wool, gently inserted, and applied at the bottom of the external auditory passage, so as to come in contact with the portion of membrana tympani which still remained. "The result was astoundingly successful. On the evening of the day this lady joined the family dinner party, and heard the conversation going on around her with a facility that appeared to all present quite miraculous."

About the year 1853 Mr. Toynbee contrived another kind of "artificial membrana tympani," consisting of a thin disc of vulcanized Indian rubber, with a stem of fine silver wire. He explained the improvement in hearing which was sometimes attained by its use, on the assumption that the tympanum must necessarily be a closed cavity. The appliance invented —or, it would scarcely be unfair to say plagiarized— by Mr. Toynbee, obtained some celebrity, and was appreciated, chiefly by Continental aurists, for a time, more highly (as I think) than its merits deserved. Meanwhile, for some years, controversy ran high between the two eminent men who thus each claimed to have originated the same truly wonderful and valuable discovery; and the war of words was long carried on with an unseemly bitterness, which was much to be regretted, on both sides. I could not give even a very concise account of the treatment I so highly esteem without some reference being necessary to the history of its introduction; but to enter into any detail of angry discussions long past would be from this place both unbecoming and distasteful. I willingly go on to consider the application of the artificial membrana tympani.

Mr. Toynbee's idea appears to have been that the

tympanum being in its normal state a chamber closed externally, any mechanical substitute for the absence or defect of the natural outer wall (the membrana tympani), must necessarily be only useful in so far as the cavity could be *reclosed* by its means. I had thought that this assumption was scarcely now adopted by any aural surgeon or physiologist; and it was therefore with some astonishment that I read, in an article on ear disease printed in the latest (this year's) edition of a work of high repute, a quotation from Toynbee, giving minute directions for fitting the Indian rubber membrane accurately to the meatus, &c., &c. The author of the article in question gives directions similar in substance, showing that he shares in Toynbee's impression, that "closure of the tympanum is the object always to be aimed at." Yet, in practice, the same writer would seem to have found that better hearing can be attained by letting in a little air, for in the next sentence he says, "I have, however, sometimes found that a portion of the membrane cut to the shape of the aperture, but a little larger in size, applied directly to the ruptured spot, has answered better than when the whole membrane has been covered." Now, as it would be obviously impossible to maintain a portion of the little vulcanized Indian rubber disc accurately pressed against the hole in the membrane, the true explanation appears to me, that in such a case sonorous vibrations were able to pass *beside* the appliance, and so to reach the labyrinth, while the remaining portion of the membrana tympani and the ossicles were supported.

My own opinions as to the mode in which Yearsley's simple and effectual contrivance acts in producing amendment in hearing, when the natural membrane is perforated or lost, have been repeatedly indicated in former pages of these lectures. Sound scientific principles and long practical experience, both point distinctly to one conclusion. The benefit is derived from support being given to the ossicula, by which they are enabled to exercise that due pressure at the fenestra ovalis, which keeps the membrane of the fenestra rotunda in a condition susceptible of vibrations, and capable of transmitting them to the nerve-expansion in the labyrinth.

Mode of applying the artificial membrana tympani.

A small piece of cotton-wool moistened with any suitable fluid is first lightly compressed into a little wad or oblong pellet, generally about half-an-inch long, and then introduced into the meatus with a pair of small forceps. Next, it is gently pushed down to the bottom of the passage so as to press upon the remaining portion of the membrana tympani. With a blunt-pointed probe, or the bar of steel at the other end of the forceps which are made specially for this purpose, the wool can be carried to any particular spot on which its touch produces the improvement in hearing. When the surgeon adjusts it, he may use the speculum, and ascertain the precise segment of the membrane, where the pressure of the artificial support appears to be the most useful. After instruction, lasting from a few days to some weeks as may be required, the patient will generally be able to apply it for himself; and it is quite astonishing with what adroitness an intelligent and painstaking patient will sometimes hit the exact spot on which to place the cotton-wool, after he has a few times had it adjusted by the surgeon. Care must be taken that the artificial membrane *does not entirely cover the opening into the tympanum*, because if it does, the sonorous undulations will not pass *beside* it on their way to the drum and the labyrinth, but *into* it, and there become quenched.

Any other known form of artificial membrana tympani will, according to my experience, not only cause irritation, and increase the purulent discharge, but will be intolerably painful even if worn only for an hour or two. I have tried spiral coils of elastic materials, horse-hair, silk-worm gut, the finest possible silver wire, &c., as substitutes, but all are too harsh and irritating to the delicate structures against which they are placed, and I have always been compelled to return to the employment of Yearsley's contrivance. Besides the direct mechanical action of this remedy, which produces such satisfactory results on deafness, the plug of wool will, by being saturated with an astringent and by its uniform gentle pressure upon the suppurating structures, frequently cure the otorrhœa. With this object, a solution of sulphate of zinc (gr. iij—v ad. ℨj.) substituted for water in moistening the wool, may sometimes be used with great advantage.

. Occasionally, the defective membrana tympani, and the whole contents of the drum, will become so permanently benefited by the application of the cotton-wool in this way, that its continued use becomes unnecessary. The purulent discharge having been cured, the mucous membranes whence it originated restored to their healthy condition, and the ossicular connections, previously relaxed, having become braced up, the conducting portions of the middle ear may resume their proper functions. The artificial membrane may now be dispensed with, for a merely existing perforation (the tympanum being otherwise healthy) will cause only slight impairment of hearing.

In the cases where there is no perforation of the membrana tympani, but a separation has occurred between any two of the ossicula (generally the incus and stapes), as instantaneous and surprising an effect upon the hearing will ensue from the pressure of a bit of wool upon the membrana tympani, as is produced by the same artificial membrane in perforations. Case No. 2, which was seen by some of you at this hospital, is recorded as an illustration of the happy results which followed the application of the moistened wool in both ears, although their respective condition differed,—the one membrana tympani being largely perforated, but the other entire; the disease having been induced by scarlatina. Lastly, when the drum-head is altogether gone, and even one or two ossicles are also absent, if only the stapes remain, and the membrana fenestræ rotundæ be free from thickening, the hearing may, as we have witnessed, be completely restored by this most simple and valuable invention of Dr. Yearsley,—a discovery (as was remarked to me lately by an eminent colleague) ranking with, or perhaps surpassing, Von Graefe's operation of Iridectomy for the cure of glaucoma in the eye, the disease which, until 1857, had remained an *opprobrium* of ophthalmic medicine.

CASES.

CASE 1.—*Membrana tympani in both ears nearly destroyed by scarlet fever more than twenty-five years previously. Hearing restored, and otorrhœa cured, by the artificial membrane.*

Miss H. W., twenty-nine years of age, consulted me in July last. Has been deaf ever since she had scarlet fever in infancy. Hears better at some times than others, which seemingly accounts for her speech being comparatively little interfered with. She mentioned the names of five aural surgeons whom she had consulted, but neither of whom had made use of any appliance to remedy the deafness. She now requires to be spoken to in a loud voice at the distance of a yard. My watch is heard only when in firm contact with the auricle. There is a most profuse and offensive creamy muco-purulent discharge from both ears. The Eustachian tubes are pervious, and a full stream of air can be driven through the tympanum by the Valsalva experiment, and rushes bubbling among the secretion out of the meatus. General health not good. After careful syringing (which produced giddiness) I inspected the ears, and found that the right membrana tympani had been so nearly destroyed, that only its rim remained; the ossicles were displaced. On the left side, the only vestige left of the membrana tympani was a small portion on the upper wall of the auditory canal. Malleus and incus lost. Interior of the tympanic cavity granular, red, and thick.

Treatment.—As soon as Yearsley's artificial membrana tympani was applied in each ear, the patient heard the striking of the metronome-bell distinctly several feet off, and my watch at twelve feet. She was desired to procure some of the cotton-wool artificial membranes with thread attached. On subsequent occasions I inserted these, saturated with zinc lotion. She was able to remove them every night and re-introduce them in the morning. At each visit to me she was instructed how to use the bit of cotton wool, and she soon learned to adjust it properly. At her second attendance the hearing distance had increased to eighteen feet on the right and fifteen feet on the left side. At the third to twenty-five feet, or the length of the room; and this degree of hearing she retained two months afterwards, when I last had an opportunity of examining her ears. The otorrhœa had entirely ceased. Three medical friends and several patients have seen this highly satisfactory case.

Case 2.—*Deafness for five years from scarlet fever. Perforation of the right membrana tympani; the left entire, but ossicles probably disconnected by the former purulent inflammation. Hearing greatly improved on* both *sides, by the artificial membrane.*

R. H., twenty-one years of age, came from Padstow,

Cornwall, to attend as an out-patient at St. Mary's Hospital, in December, 1869. Became deaf five years ago, when she had scarlatina. Otorrhœa from right ear ever since; has never been aware of any discharge from the left. Watch only heard in contact with either auricle. Right Eustachian tube pervious, left, impervious to the Valsalva experiment, but was rendered patent by the Politzer bag. Right membrana tympani perforated in the centre, through which issued a shreddy mucous secretion. The usual "perforation-murmur" was produced on inflation. Left membrane was observed to be whole, but flattened and opaque; tubercle or malleus prominent. The handle, owing to interstitial deposition, could not be seen. As soon as a piece of moistened cotton-wool was introduced and properly adjusted on the right side, the hearing rose to three feet. On the left side no improvement resulted from any treatment adopted, until it occurred to me, after several attendances, to press a small bit of cotton-wool firmly down upon the membrana tympani, when a surprising increase of hearing-power was given—the ticks of the watch being distinguished at a distance of four feet, and conversation heard easily at any part of the room. I was led to try the effect of artificial support, by a circumstance the patient related, that on one occasion when the left ear had been filled with water, considerable improvement in hearing continued as long as the fluid was retained there. I thence inferred that *pressure* upon the membrane was required to support the ossicles, which had probably been disconnected during the bygone catarrhal process, but without erosion of the membrana tympani having taken place. Subsequent trials of the cotton-wool plug never failed to effect the same amount of improvement in the hearing, and this patient returned home after a month's attendance, "quite a different person;" and hearing on both sides sufficiently well for all necessary purposes, whenever she had in the morning adjusted the artificial supports.

Instances more or less resembling the foregoing might easily be cited in almost any number; but these, with the references to some others which have been made incidentally in previous lectures, must suffice. I am, for many reasons, very averse to the needless multiplication of cases; and I consider that the desired object of impressing the listener's or reader's mind is best attained by choosing carefully only a very few typical ones as illustrations—which can be thoroughly examined and made your own without burdening the memory.

Conclusion.

I ought, perhaps, to apologise for the length to which these lectures have extended; but I can only say I have endeavoured to condense them as much as was possible, consistently with an approach to a fair treatment of the subject. If I could venture to hope they would be read with any spark of the intense interest and delight which I have felt in drawing them up, I should feel confident that their author's tediousness would be forgiven.

12

INDEX.

Air-douche, the, 91
Anchylosis of stapes, 142
Anatomy of Eustachian tube, 63
    ,, of membrana tympani, 40
    ,, of the tympanum, 52
Artificial membrana tympani (Yearsley's), 261
    ,, ,, ,, (Toynbee's), 263
Aural catarrh, 99
        acute, 101
         ,, treatment of, 121
        chronic, 137
         ,, diagnosis of, 175
         ,, treatment of, 189
        purulent, 222
            pathology of, 225
            treatment of, 245
  ,, ,, ,,
  ,, surgery, neglect of, 3
Auscultation of the ear, 71

Catarrh, aural, 99
    ,, acute aural, 101
    ,, chronic aural, 137
    ,, naso-pharyngeal, 156
    ,, purulent aural, 222

Catheter, Eustachian, 85
Catheterism of Eustachian tube, 83
      ,, effects of, 92
      ,, history of, 84
      ,, precautions to be observed in, 88
      ,, compared with the Politzer method, 82
Cavity of tympanum (see tympanum)
"Cone of Light," the, 45

Deafness, nervous, how distinguishable, 38
    ,, degree of, how ascertained, 26
Deaf-muteism, 5, 228
Deglutition, effect of on Eustachian tube, 65, 77

Ear, the divisions of, 7
Eustachian catheter, 85
        tube, anatomy of, 63
    ,,   ,, catheterism of, 83
    ,,   ,, in acute catarrh, 113
    ,,   ,, in chronic catarrh, 150
        ,, obstructed, evidences of, 177
    ,,   ,, uses of, 64
Examination, aural, methods of, 21

Facial paralysis, 226
Fauces in acute catarrhal inflammation, 114
  ,,   ,, chronic catarrhal inflammation, 155
Fenestræ, the two, importance of, 57

Gruber, Dr., on injecting the tympanum, 211

Hinton, Mr., on mucous accumulations, 148
    ,,   ,, on tinnitus, 171
    ,,   ,, on the treatment of otorrhœa, 258

Illumination of ear, modes of, 22
Inspection, ocular, importance of, 10

INDEX.

Jago, Dr., his investigations, 151
  ,, ,, on tinnitus aurium, 169

Listening, physiology of, 27

Manometer, the, 67
Mastoid cells, chronic inflammation of, 160
    ,, functions of, 161
    process, in otitis, 251
  ,, ,, treatment of abscess of, 252
Meatus externus, examination of, 21
    ,, in acute catarrh, 111
  ,, ,, polypi in, 244
Membrana tympani, anatomy and appearances of the healthy, 40
    effects of catheterism upon, 97
    emphysema of, 96, 148
  ,, methods of examining, 21
    in acute catarrh, 106
    in chronic catarrh, 143
    in purulent catarrh, 233
    observations on puncturing or incision of, 206
    perforation of, 237
  ,, the artificial (Yearsley's), 261
    (Toynbee's), 263
  ,, ,, ,, ,, mode of applying, 265
Meyer, Dr., on adenoid growths, 156
Myringitis, 107

Naso-pharyngeal catarrh, 156
Noise, hearing better in, physiological explanation of, 29

Ossicula auditûs, 54
  ,, ,, muscles of, 55
  ,, ,, functions of, 56
Otitis, see aural catarrh, purulent.

Otorrhœa, 235
　　,, treatment of, 252
Otoscope, 72

Perforation of membrana tympani, 237
　　　　　　diagnosis of, 240
　　　　　　treatment of, 252
　　,, artificial membrane in, 265
Politzer's method of inflating the ear, 77
　　,,　　　,, compared with others, 80 and 82
　　,, bag, the author's improvement upon, 78
Polypi, aural, 243
　　,,　　,, treatment of, 252
Physiology of listening, 27
　　　　　　hearing during noises, 29

Specula, aural, 21
Speculum, pneumatic, 147
Stapedius muscle, functions of, 56, 183
Stapes, anchylosis of, 142
　　,, impaction of, 28, 142
　　,, diminished mobility of, 185

Tensor tympani muscle, functions of, 55, 183
Throat in acute catarrh, 114
　　,, in chronic catarrh, 173
Tinnitus aurium, 164
Tonsils, diseases of, 155, 113
Toynbee, Mr., his investigations, 3, and elsewhere
　　,,　　,, his artificial membrane, 263
Tuning-fork, phenomena and uses of, 36
Tympanum, anatomy of, 52
　　,, in acute catarrh, 101
　　,, in chronic catarrh, 138
　　,, in purulent catarrh, 222

Tympanum, different methods of inflating and examining the, 70
,, mucous accumulation in, 140, 148, 205
,, injections into, 209, 211

Valsalva method of inflation, 74
,, ,, ,, compared with Politzer's, 80
Von Tröltsch, Dr., his excellent Treatise on the Ear, 64
,, ,, on "A Great Gap in our Knowledge," 180
,, ,, on "Incision of the Membrana Tympani," 210

Wilde, Sir Wm., on "Cone of Light," 45
,, ,, on Eustachian Tube, 151
,, ,, on Tinnitus Aurium, 164

Yearsley, Dr., on "Artificial Membrana Tympani," 261.

www.ingramcontent.com/pod-product-compliance
Lightning Source LLC
Chambersburg PA
CBHW031249250426
43672CB00029BA/1388